Cover: Two unidentified miners pose for a photographer as they relax on the covered porch of their cabin near the Mizpah Mine in the Hoodoo Mining District.
(Latah County Historical Society (25-4-8)

Gold Mining in the Hoodoo Mountains of North Idaho, 1860-1950

Grubstaking the Palouse

Gold Mining in the Hoodoo Mountains of North Idaho, 1860-1950

by Richard C. Waldbauer

Washington State University Press
Pullman, Washington

Latah County Historical Society
Moscow, Idaho
and
Whitman County Historical Society
Colfax, Washington
1986

Washington State University Press
Pullman, Washington

Copyright 1986 by the Board of Regents of Washington State University

All Rights Reserved. No part of this book may be reproduced or transmitted in any form or by any means, electronic or mechanical, including recording, photocopying, or by any information storage and retrieval system, without permission in writing from the publisher.

Printed and bound in the United States of America
Washington State University Press
Pullman, Washington 99164-5910

Grubstaking the Palouse

ISBN: 0-87422-021-1 (softcover)

1 2 3 4 5 6 7 8 9 10

Acknowledgments

The importance of interpreting local history lies in what anthropologist Clifford Geertz has called "little universes of meaning." That is, we gain a better perspective of the human experience by understanding how and why people interact within their particular social contexts. Local histories record past behavior, but they are also valuable for understanding the social world. I can immodestly extol the virtues of local history because the initiative and cooperation which brought this one into publication came from folks long-committed to and well-versed in the contributions of that kind of history.

This project was an exploration into archaeology, historical documentation, and ethnography. The Hoodoo Mining District is one of the landmark treasures of the Palouse because of the stewardship provided by the Palouse Ranger District of the Clearwater National Forest. While the archaeological project was underway, the Resources staff was headed by Jim Dewey, the Timber Management staff by Blake Ballard, and the Ranger was Dave Colclough. They actively supported the project's fieldwork.

Essential contributions to the ethnographic project were made by Frank Milbert of Potlatch and Roy Chatters of Pullman. The oral history collections of the Latah and Whitman county historical societies were extensively consulted. Both collections represent systematic efforts to record the qualitative remembrances of people who have lived in the Palouse. In the case of the large Latah County collection, many of the interviews have been transcribed for easier comparative analysis.

The documentary project received the support of the staffs of the Latah, Nez Perce, and Whitman county offices. Essential were the local history resources found in the Whitman (WCHS) and Latah (LCHS) county historical societies. I want to thank Walter Nelson, president of WCHS for his support of the project; Thomas Sanders, director of the Washington State University Press for recognizing the value of regional historical publication and for cooperating with local historical societies in efforts such as this; Dorothy Clanton, at the time this manuscript was considered, chair of the LCHS publications committee who provided strong support and reviewed the work; Mary Banks, LCHS publications committee member who retyped the manuscript prior to publication; Sharon White of the University Press who designed the book; and J.D. Britton of the University Press for editorial assistance.

Fred Bohm, managing editor of the University Press as well as a WCHS trustee, assisted greatly in both of those capacities. I am indebted to him for his commitment to the joint publication project, and for his editorial skill. Mary Reed, director of LCHS, helped to gather the illustrations used in the book.

Finally, I express appreciation for the contributions of Keith Petersen, who helped as editor, photograph collector, and friend. As colleagues in research and fellow employees at LCHS, we have explored many aspects of the interpretation of local history. His hard work has been critical in the process of publication, and his inspiration has been a constant guide in creating this history of the Palouse River Valley.

Richard C. Waldbauer
Washington, D.C.
April 1986

Contents

Introduction

An unidentified Latah County miner poses with a shovel in front of his cabin. In the foreground can be seen the remains of a case of Calumet Corn Starch, an empty tin of cooking oil or shortening, and a supply of wood for the cook stove. *(Latah County Historical Society, number 25-4-12)*

Few of America's laws protect the rights of the individual better than the United States Mining Laws of 1866, 1870, and 1872. It might be said that those statutes rank with the fundamental guarantees of the American brand of liberty—the preference for equalization of economic opportunity before freedom of social action. In addition, few of America's laws have survived so long with so few revisions. There were no significant influences on federal mining law until the

early 1970s, when the National Environmental Protection Act and its associated legislation required the filing of environmental impact statements before development of minerals claims could begin. Even then, the individual placer miner was insulated from those restrictions so long as he/she used equipment capable of moving less than one cubic yard of gravel per day.

The great significance of these laws from a cultural perspective is that they evolved from the social experience of communities of miners. The first statute, passed in 1866 with the help of Nevadan representatives promoting the interests of the Comstock Lode, had three major features. Two had to do with when and how miners could go onto the public domain and exploit minerals. The third specified that local arrangements of the miners themselves were to be respected and continued.

> The preamble (to the law) prescribed that the lands were to be open "subject to such regulations as may be prescribed by law, and subject also to the local customs or rules of miners in the several mining districts." Again, within general limits "the local laws, customs, and rules of miners" were to be controlling when the federal government undertook to accept and survey claims (Rodman W. Paul, *Mining Frontiers of the Far West, 1848-1880.* Albuquerque: University of New Mexico, 1963).

The key to these guarantees was the folk/moot phenomenon known as the mining district. When that law refers to "local customs" and "rules of miners," it means the system of self-government miners devised for themselves during regular meetings of the mining district. This was an extraordinary recognition of an institution that could be said to have been less than 20 years old at the time, because it rose from the gold rush experience of 1849. Further, it was an institution that had not undergone rigorous examination in law as a duly-constituted representative assembly. It could be argued that miners' meetings were little more than vigilante committees. On the other hand, it could be also argued that the meetings were examples of the best traditions in American representative democracy.

Individual rights were the paramount concern in any action decreed by majority vote in miners' meetings. Most miners were simply trying to attend to their own business and leave others' alone. As a result, anything tending to prevent all individuals from having an equal chance to mine the riches in their district was accorded a rapid hearing and visited with quick justice. Miners could not squander valuable time during the season to delve deeply into the legal ambiguities of a particular situation.

The story of the Hoodoo Mining District is primarily about how important a small, local gold placer mining area was to the economic development of a much larger region. It is an attempt to show that for every bonanza, boom-and-bust, hell-bent-for-leather mining camp on the frontier there were dozens of quieter backwaters where the contributions to settlement were less notorious and more enterprising. Even if the individual miners themselves did not stay permanently in the Palouse, the fruits of their energy fed others with visions of farms, factories, churches, and schools.

That is the surface interpretation. The subtle meaning is in the vein of community relations. To mine this, it is necessary to understand that the wealth of the Hoodoos was social as well as mineral. The miners made up what has become a hallmark of American cultural history—a community in the wilderness. They could do this

Paul Bockmier's family standing in front of their home in Fisher's Addition, east of Palouse.
(Whitman County Historical Society, Paul Bockmier Collection)

because of the nature of their unique institution, the mining district.

Miners came and went. As we shall see over the course of the history of the Hoodoo Mining District, people followed the lure of gold for a variety of reasons. However, long after any one person was gone, the institution of the mining district remained. By the end of the period, the very rules that protected the individual miners also served to protect that great mythical individual—the corporation. It is a fair question to ask whether it was technological advantage or legal fiction that drove placer miners from their claims in the face of competition from the dredgers, bulldozers, steam shovels, and conveyor belts of corporate enterprise. Mining companies brought more than just engineering skill and economy of scale to minerals regions. They were accorded legal protection that had been originally devised for humans.

But it is people who engage in enterprise, and the pioneer communities of the Palouse River Valley have changed dramatically into those we know today. The role of the Hoodoo Mining District in those changes is no small part of the story.

"Source of the Peluse," a lithograph delineated by the prominent early artist of the Pacific Northwest, John Mix Stanley, was based on a drawing by Gustavus Sohon, an equally important artist who accompanied the Isaac Stevens Party in the mid-1850s. The illustration was included as a plate in the Reports of Explorations and Surveys to Ascertain the Most Practicable and Economic Route for a Railroad from the Mississippi River to the Pacific Ocean, *commonly called the* Pacific Railroad Reports, *published between 1855 and 1861. The Stanley/Sohon lithograph is among the earliest illustrations of the Palouse hills. The view is from the top of Kamiak Butte in Whitman County, Washington, looking east into Idaho and the Hoodoo Mountains.*

(Private Collection)

Gold!
Where They Found It

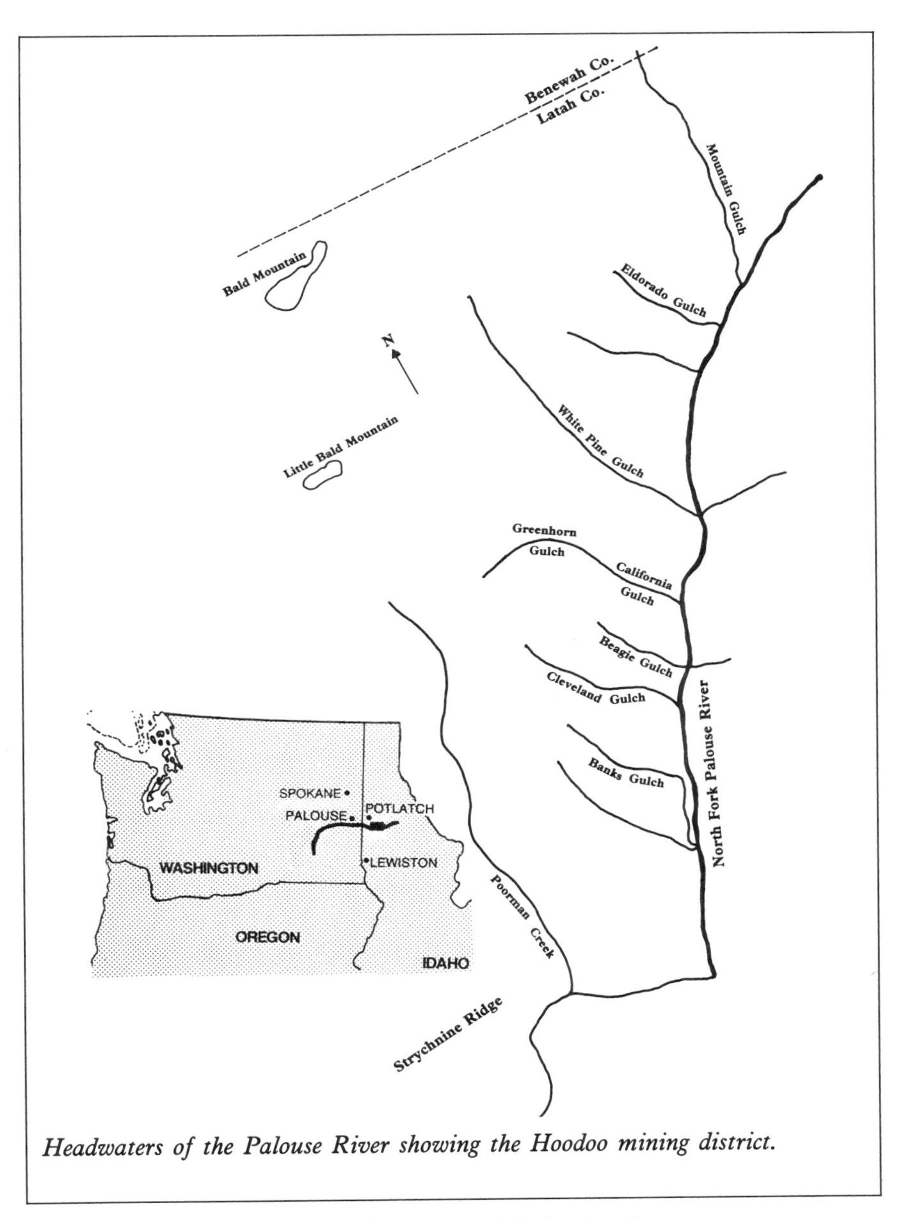

Headwaters of the Palouse River showing the Hoodoo mining district.

Gold rushes often provided the impetus to mass settlement by whites in the Mountain West. The legendary life of the gold fields played an important role in the development of San Francisco, Seattle, Denver, and Portland. But there

Taken in 1882, this is one of the earliest extant photographs of the Town of Palouse. The structure to the far left (behind the pine tree) is William Breeding's flour mill; the large two-story building at the top of the hill is O. E. Clough's hotel; the dwelling in the foreground was probably a house of prostitution. (Whitman County Historical Society, Paul Bockmier Collection)

were also numerous, smaller and less significant gold discoveries with limited national impact. The Hoodoo Mining District in northwestern Latah County, Idaho was one of these. Though it is only rarely mentioned in histories of the Pacific Northwest, the permanent settlement of the upper Palouse River Valley owed much to the success or failure of the Hoodoo miners.

This history covers the period from the mid-1850s, when the upper Palouse River Valley was first mentioned in documentary sources, to the early 1950s, when the last major effort to placer mine the Hoodoos took place. During that time, several gold strikes were made, and a significant amount of mineral wealth was recovered. Initially, the Hoodoo Mining District was a focus for the development of settlement because of its very nature as an organized community. The gold nuggets that miners exchanged for foodstuffs and supplies helped establish a flow of currency among farmers and entrepreneurs in the area. The wealth that was distributed, however, could have meaning only to people determined to build communities of farms and towns. Therefore, the history that follows is less about the gold that was discovered and more about the lives of people who made use of it.

The Hoodoo Mining District is located on the headwaters of the Palouse River where it flows out of the Hoodoo Mountains in northwestern Latah County, Idaho. The specific location for the legal boundaries of the mining district have fluctuated, but it has always centered on four gulches that drain into the North Fork of the

Miners pose in front of ore sacks at the Mizpah Mine in 1919.
(Latah County Historical Society, number 25-4-7)

Palouse River: the Hoodoo, Greenhorn, California, and White Pine Gulches. These and all the important placer-mined drainages flow into the North Fork of the Palouse River from the north and west. Other important tributaries to these rivers are Mountain Gulch, Eldorado Gulch, Moscow Gulch, Cleveland Gulch, Banks Gulch, Poorman Creek, Strychnine Creek, and Excavation Gulch.[1] The only major hardrock mine is located just east of these drainages, across the Baby Grand Mountain divide on Mizpah Creek.

The Palouse River meanders westward from Latah County into Whitman County, Washington and eventually empties into the Snake River. In its upper reaches the river emerges from the Clearwater Mountains, of which the Hoodoo Mountains are a part, and passes through a broad valley. The Hoodoo Mining District is located at the upper end of the valley. The present site of Laird Park, about four miles northeast of Harvard, Idaho, and Strychnine Ridge mark the end of the bottomland. From here the river passes through the towns of Harvard, Princeton, Hampton, and Potlatch. Its major tributaries in this section are four creeks known as Meadow, Jerome, Hatter, and Gold. On the north the valley is shadowed by Gold Hill, a 4,700 foot isolated promontory, and to the south it is bounded by the Palouse Range. Moscow Mountain, at about 5,000 feet, Bald Butte, Basalt Hill, Mt. Margaret, and Mica Mountain are the most significant points in that Range.

Manuscript map of the reconnaissance of Pyramid Peak (Steptoe Butte) during the Mullan Road exploration by Theordore Kolecki, topographer, 1859. Existing trails and timber were shown along the portion of the Palouse River above the military road crossing. (Cartographic and Architectural Branch, National Archives and Records Service)

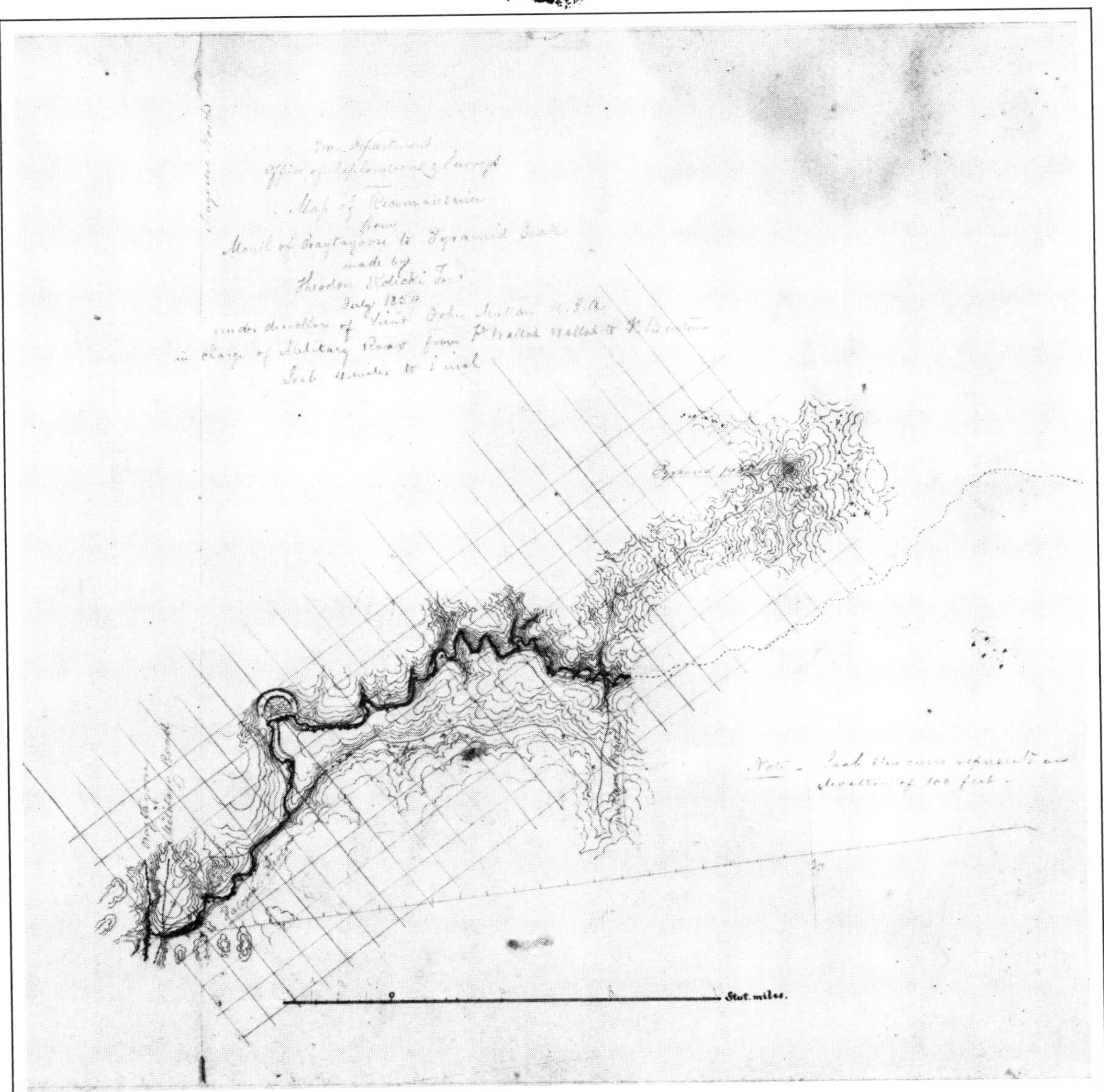

Manuscript map of the reconnaissance from the upper Palouse River and the area around what is now Moscow, Idaho, during the Mullan Road exploration, by Gustavus Sohon, surveyor, 1859. The dates shown indicate the explorers' campsites as they passed through the camas prairie to the heights east of Moscow. (Cartographic and Architectural Branch, National Archives and Records Service)

Flowing west, the Palouse River leaves the valley and flows into the open country of the Palouse Hills. Two miles west of Potlatch, near the vicinity where U.S. Highway 95 now crosses the river, is Kennedy Ford, once known as Palouse Bridge and the only suitable crossing for horses and wagons along the flanks of the mountainous country.[2] Seven miles downstream, the town of Palouse, Washington straddles the Palouse River. The relationship between this country and the mining district—from Palouse to the Hoodoos—played a significant role in the economic

Palouse, Washington just prior to the "great fire" of 1888. At the left end of the bridge stood the town grist mill. Above the grist mill, along the hill, was the road to Almota; to the right of the bridge was "New Town" and the road to the Hoodoo Mining District. (Whitman County Historical Society, Paul Bockmier Collection)

geography and historical development of the Palouse River Valley.

The Hoodoo Mountains are geologically different from the rolling Palouse Hills to the west. The mountains are characterized by enormous folds and faults that created terrain laced with high, massive divides and deep valleys. The placer mines were located between 2,600 and 4,600 feet above sea level. The streams that drain the steep mountains scoured vast amounts of earth, sand, and gravel; much of this material has been deposited along the stream bottoms as alluvial sediment which is often several miles long and as much as twenty-five feet deep.[3]

The power of falling water was an essential feature of the Hoodoo Mining District. First, it was erosion of the mountains by water that laid the gold in placer deposits along gulches and streams. Then, the miners harnessed the power of falling water to expose those deposits and retrieve the gold. They had to have precise knowledge about drainage systems because the water flow was seasonal and claims could not be worked without it.[4] For example, attempts to work claims in the early spring presented problems. The narrow gulch bottoms and west to north-facing slopes frequently were deep with snow at that time, and it was not unusual for

This view of Palouse, looking to the southwest, shows the devastation caused by the 1888 fire. Like many frontier communities, Palouse was constructed largely of wood and was highly combustible. Lacking adequate means to combat fire, frontier cities such as Seattle and Spokane, and towns such as Colfax and Pullman, all suffered fates similar to Palouse. (Whitman County Historical Society, Paul Bockmier Collection)

them to remain so well into May. By mid-March, however, ridge crests, major divides, and south to southeast-facing slopes began to clear. Hillside and high gulch placers fed by ditch-transported water could be worked.[5] The best miners took full advantage of the season's water by locating ditches close to the snow line to catch the early runoff, but not so close as to risk being blocked by frozen drifts.

Geologists have not identified the source of Hoodoo gold. However, they have made some general assessments about "mineralization," the geologic process by which gold was concentrated.[6] These analyses suggest that the enormous events of mountains folding and faulting were accompanied by flows of molten rock and pure metal which found their way into cracks and crevices as veins of mineral ore. The pattern of radiating faults discovered at the head of White Pine Gulch is typical of places where gold may have accumulated. Later, many fault lines themselves became paths by which flowing water eroded the surface into streambeds. Mountain Gulch, California Gulch, and Banks Gulch are examples of fault lines superimposed with streambeds.

There is one mystery in this entire sequence of events. Between the two geologic formations in which gold is thought to have accumulated, a certain type of rock is missing. Such absences are called "disconformities" by geologists, and they indicate that important parts of the geologic history cannot be told. For the Hoodoo Mountains, the disconformity means there is no way of knowing how the gold-bearing formations were exposed to the erosion that eventually created

placer deposits. This was a source of some confusion among the Hoodoo miners as well, and many explored their hillside placers as if they were ancient streambeds. Their hope was to strike the bonanza deposits from which came the "showings" they had found in the gulches below.

But despite the disconformity, the geologic criteria for placer minerals were met. Erosion had cut deeply into the mineralized strata to create major alluvial deposits along the Palouse River, Poorman Creek, White Pine Gulch, and Mountain Gulch, as well as significant alluvial beds on the west side of the North Fork of the Palouse River.[7]

Mines and Community Growth

A building sometimes described as the "post office at Carrico" was supposedly constructed in 1862. Although no records have been found to show that the government ever authorized a post office at Carrico, the structure might well have served as an unofficial mail distribution point. The Community of Carrico was founded in the 1880s by Jerome Carrico and, at times, more than 250 miners operated in the vicinity of this location at the head of Gold Creek. The photo, taken in 1927, shows Paul Bockmier, Senior (left) with Spokane businessman W. A. Johnston standing in front of the structure's remains. (Whitman County Historical Society, Paul Bockmier Collection)

When early California placers could no longer be worked by simple methods, many miners moved north. These were men who had gone west in the rush of 1849 and learned placer mining through the use of a pan, rocker and sluice. They mined their claims individually or in small associations with a few of their neighbors. By cooperative effort, a company of placer miners considerably increased the quantity of gravel they processed.[8]

Three unidentified miners pose in a shaft of the Copper Chief Mine.
(Latah County Historical Society, number 25-4-3)

After 1851 more sophisticated methods and capital were needed to exploit California gold fields, and discoveries at the Rogue River in Oregon started the northward rush by prospectors. By 1855, the movement reached Fort Colville, north of Spokane on the east bank of the Columbia River in Washington Territory.[9] The coming of miners was a hopeful sign for the economies of Oregon's Willamette Valley and Washington's Puget Sound regions where business had been stagnant since 1852. There was general hope that an influx of currency from the sale of miners' supplies would relieve the depression.[10]

Development of the Colville area was not uncontested. The Indian populations became alarmed by the stampede of miners. Treaties with Columbia Plateau-dwelling peoples initiated by

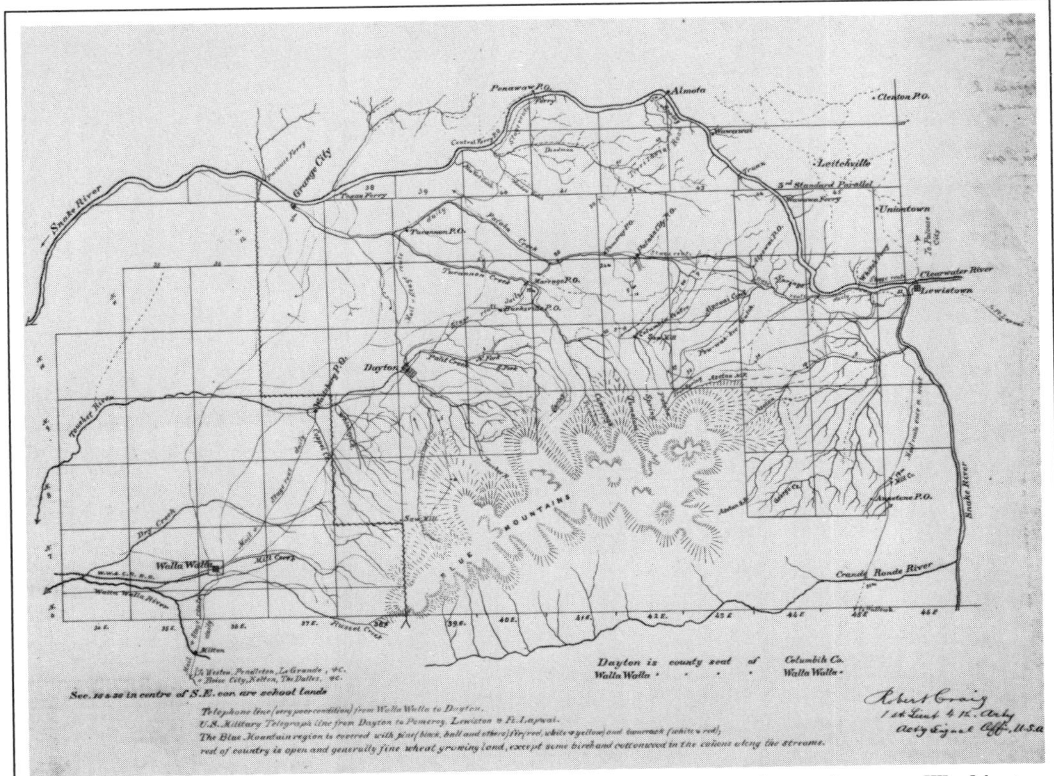

United States Army Signal Corps map of the mail and stage routes in southeastern Washington Territory in 1879. The territorial road between Pomeroy and Almota was shown to have been little more than a trail. (Cartographic and Architectural Branch, National Archives and Records Service)

Isaac I. Stevens in 1855 failed to satisfy either Indians or whites, and miners en route to Colville across the Yakima country were attacked. By that time, the militia called out by the Oregon and Washington territorial governors began to roam the country. General John E. Wool, commander of the regular army's Department of the Pacific, claimed that "the war had been precipitated by the treaties of Stevens, that the volunteers had entered it largely in order to plunder the Indians, and that citizen speculators had fostered it for the purpose of getting more money into the country from the Federal Government."[11]

This unrest stunted mining activity around Fort Colville through 1859, causing gold rushers to disperse to the Okanogan country and the Thompson River in British Columbia. Gold discoveries along the Fraser River sparked another rush of miners, especially from California via Victoria, British Columbia. Interest in the Colville region was revived in 1859 when the regular army built forts and provided large escorts for the United States Boundary Commission, whose representatives were charged with marking the United States-Canada boundary.[12]

The final demonstration of uncontrolled intrusion by whites onto the Columbia-Plateau was the reversal of General Wool's policies. The knowledge that it would be impossible to prevent the advance of miners and settlers caused General Newman S. Clarke, Wool's successor, to allow

Enduring the economic dislocation caused by the Panic of 1893, the town of Palouse entered an era of dramatic growth. In the upper photograph, the business district is shown as it appeared in 1896. Ten years later in 1906, the impact of mining, lumbering, and agricultural activity on the community was even more pronounced (see lower photograph). But undercutting of the town's fledgling lumbering industry by Frederick Weyerhaeuser's mill at Potlatch, Idaho; the failure of miners to discover major ore deposits in the Hoodoos; and the impact of mechanization on regional agriculture would shortly drive the community into decline. (Whitman County Historical Society)

access to the region and recommend confirmation of Stevens's treaties.[13]

Thus, by the time of the Mullan Road survey and prior to major gold discoveries in northern Idaho, prospectors and adventurers were present in significant numbers, wandering over the entire region. E.D. Pierce, whose discoveries precipitated a massive rush to the Nez Perce mines, prospected the territory intermittently prior to leading a party to Oro Fino Creek via a route through present-day Latah County. Also, by this time the nature of prospecting parties was clearly established. They usually consisted of from five or six men to as many as fifty in a single group. They were organized around a core of experienced miners who had usually done apprenticeships in the California fields. An expedition that discovered valuable deposits ensured success by immediately organizing a mining district, again, on the California style.[14]

At the time of the earliest descriptions of the upper Palouse River country in Latah County, gold prospectors were consistently visiting remote regions in northern Idaho. When John Mullan reported his findings on the military road survey of 1860, it is clear that he noted various independent reports which came to him. "At the headwaters of the (Palouse River) and its tributaries," he stated, "limestone is said to be found, and there also, in places, the soil is fertile, and lying as it does under the slopes of the mountains, and in close proximity to the Nez Perce mines, it is not at all improbable that the grazier and agriculturalist will find at no distant day tracts of land that will amply repay their reclamation."[15] By mentioning the supposed occurrence of limestone on the upper Palouse, he probably referred to prospectors' observations. It was known that primary ore deposits can be found as mineralized chlorides upon the contact zone of limestone.[16] Since this type of deposit was readily smelted in crude furnaces, such occurrences would have been of interest for immediate regional economic development.

Two subsidiary parties of Mullan's Military Road Expedition made explorations away from the main route toward the head of the Palouse River. The party led by Gustavus Sohon, guide and interpreter, reached the north peak of the "Tat-hu-nah" Hills at the headwaters of the Smakodle Creek (South Fork Palouse River).[17] He described the country northeast and east as densely forested with pine. To the southeast were broken flats free of timber.

Sohon's mission was to explore a possible road to the Hell's Gate defile. When he first inquired of Coeur d'Alene Indians the likelihood of finding a route, they extolled the passage through the "gate," two miles north of the Tat-hu-nah Hills. They said it could be reached by following the Mo-ho-lis-sah River (main branch of the Palouse above its forks at present-day Colfax, Washington). Later, when they discovered the actual purpose for Sohon's survey, the guides refused to assist any further and advised that the road be built in Pend Oreille country:

> (Yah-moh-moh) made an energetic speech, declared his friendship for Whites, &c., but described the mountains as formidable, the forests and underbrush as impenetrable, and the streams as dangerous, if not impassable, &c., and implored me not to think of exploring the route—that if I did I would perish, and rumor would say that the Indians had killed me.[18]

The second exploratory party of Mullan's expedition was led by topographer Theodore Kolecki. He reached Pyramid Peak (Steptoe

Local businesses, such as the Ankcorn Hardware Store, profited from their retail trade with miners in the Hoodoos. Fred Ankcorn, who originally came to Palouse to work as a tinsmith, established this enterprise in the 1890s. (Whitman County Historical Society)

Butte), and from there he was able to provide the following description: "The spurs of the Bitter Root mountains, from which it (the Palouse River) proceeds were gently sloping and densely wooded. Pine timber, in scattered groves, reaches from them to within four or five miles of Pyramid Peak." During his survey for a transcontinental railroad route, Isaac I. Stevens had also described the abundance of timber in the Palouse River drainage. "We had a view from the Palouse for some thirty or forty miles," he wrote, "and the timber was apparently as large and abundant at the lower end of the valley as at our present camp" on the ridge southeast of present-day Moscow.[19]

These are the earliest descriptions of the country confronted by prospectors who ranged into the Palouse River Valley. It was wild, but the surveyors believed they found great economic potential of the kind that prospectors were searching for. There is no record of what the prospectors themselves thought of the country, and they were not even required to make a formal record of their claims until after 1866. By then the first rush to the Hoodoos had already begun.

According to local tradition, gold was discovered in the Hoodoo Mining District in the early 1860s, when a man named Hoteling supposedly discovered the metal in Hoodoo Gulch.

The Palouse business district about 1915.
(Whitman County Historical Society, Roy Chatters Collection)

However, no documentation has been found to substantiate this story. The earliest documentary evidence for this man "Hoteling" dates from 1873 when John Hoteling, John I. Doyle, William H. Flake and Ed Pearcy claimed water rights for mining on the north fork of the St. Joseph River's south fork. Their filing read, in part, ". . .we hereby claim the right to erect dams ditches flumes and any and all necessary constructions or improvements for divertion [sic] of the water of said stream to purposes aforesaid. . . ." Stories told to long-time Gold Hill miner Frank Milbert, during mining district meetings, also described the return to Gold Creek by its discoverer in about 1870. Much later, in July 1908, John Hoteling, Randall Kemp, and William Goldthwaite located claims in Greenhorn and Hoodoo Gulches. In other words, the evidence about John Hoteling's subsequent activity does not conflict with his possible discovery of gold in the area in the 1860s.[20]

Other secondary sources, based largely on interviews with early Hoodoo miners, emphasize the ephemeral nature of information relating to that period. One such interview suggests that placer gold was discovered on the North Fork of the Palouse River by Frank Points in 1872. Interviews done between 1885 and 1896 by Paul Bockmier, Sr. place the first discoveries on Gold Creek and Gold Hill in 1862. As a long-time Palouse, Washington resident and mining investor, Bockmier stated that the first pioneers in the area branched out from Pierce City and Florence in the early 1860s and came north to the Hoodoo region. This assessment supports the established conclusion that the Hoodoo mines were the ". . . base of original white activity in the Palouse."[21]

It is clear, however, that "showings" in placer gold deposits on the upper Palouse River were not sufficient to keep eager fortune-seekers there permanently. The Hoodoo Mining District, then considered to be the four gulches emptying into the upper Palouse River, reportedly could pro-

The entrance to a typical Latah County hardrock mine, this one the Copper Chief Mine in the Moscow Mountains. (*Latah County Historical Society, number 25-4-9*)

vide enough gold to pay $20 to $100 per man per day. These wages did not last long, however, and by the late 1860s profits were said to be sporadic. One of the major factors preventing profitability was the considerable expense of transporting supplies. The long haul from Walla Walla or Lewiston made only the richest properties economical to mine. Though traders, trappers, and farmers could live off the land, miners could not. They were almost totally dependent upon outside sources for food, clothing, and tools of the trade. Since Lewiston, 80 miles away, was the nearest trading point, freight rates for provisions and tools packed into the Hoodoos on ponies prevented working mines that paid less than $20 per man per day.[22]

A second factor helping to diminish the early excitement in the upper Palouse was the news of rich strikes elsewhere. By 1863 there were supposedly 100 miners on Gold Creek and a village, complete with store and post office. A short time later, the community was nearly deserted. Work on the Hoodoo mines ended with news of strikes in Alder Gulch, Montana.[23]

Further evidence of a decline in mining activity on the Palouse River is found in the field notes of General Land Office survey maps. When surveyor Henry Meldrum arrived in the summer of 1871, he captioned the river course on his map: "Placer gold mining on the Palouse River." When the surveys were completed in 1879, a note was added: "Hon. S.S. Fenn, delegate from Idaho, says there has been and still are some very rich mines on the Palouse . . . but the best have been worked out."[24]

If early gold discoveries in the Hoodoo District commanded any excitement, it meant little in terms of attracting a permanent population or sparking entrepreneurial drives by the miners themselves. For example, the first settler near Palouse City, the community that grew up just to the west of the Hoodoo District, did not arrive until 1869 when William Ewing started a ranch with 400 cattle. Ewing established his enterprise about two-and-a-half miles above the future townsite at Palouse Bridge. A man known only as "Atwood" was in partnership with Ewing, but it is not known whether Joseph Knight, Joseph Hammer, or A. Towner, who all arrived at that time, participated with them in a cooperative cattle business. It is unlikely that early cattlemen were permanent settlers because simply grazing cattle in a given area to supply the military or mining camps was a common practice.[25]

Ironically, permanent settlers were attracted to the Palouse because of a land rush at Walla Walla. Development of the Pierce and Oro Fino mines created a valuable produce market there, and as the good land around town was taken, new settlers headed northeast.

News of preparations for a Northern Pacific Railroad route further encouraged settlement along upper Union Flat Creek in 1869. By 1871, the population between Union Flat Creek and Palouse Forks (present-day Colfax) stabilized at about 200. James Perkins built a sawmill at Palouse Forks, and Anderson Cox constructed a flour mill. The area was then a part of Stevens County, with the county seat far north at Colville. Increasing settlement and political pressure chiefly from Cox and Perkins influenced the territorial legislature to create Whitman County with its seat at Colfax. Along with Walla Walla, Spokane, and Stevens Counties, Whitman County was specifically organized to exploit resources at the behest of mining and stock-raising interests.[26]

Other entrepreneurial activity followed the founding of Colfax. In 1873 James "Modoc" Smith settled at the future site of Palouse City. By 1874 settlers moved into the Deep Creek area across the line into Idaho Territory. Daniel Notman built at the Freeze Church site, Arthur Green settled on Gold Creek, and others located at Palouse Bridge, Cedar Creek, and Four Mile (Viola). A regular mail drop was established at Ewing's ranch in 1873, but was moved to Four Mile in 1874 because many of Ewing's cattle died during the winter and he left the country.[27] Immigrants moved into the Washington-Idaho border area from the north as well. Their supply routes were from Spokane Falls and Colville via Pine Creek. George W. Truax located his homestead on the site of Farmington, about 23 miles northeast of Colfax, in 1871. He became town proprietor after constructing a trading post in 1877, but until then he and his neighbors were regarded as "stock raisers and general agriculturists."[28]

It is not clear whether the placer mines of the Hoodoo District were worked continuously from the time of their discovery or if the influx of settlers during the early 1870s prompted renewed interest. It seems likely that there was scattered occupation of the upper Palouse Valley, and the most frequently recorded name in connection with the region at that time is Jim Lockridge, also called "Long Jim." For years, his cabin at the mouth of Jerome Creek, later known as Chambers Flat, was said to be the oldest structure in the vicinity. Lockridge's squatter's claim, between Gold Hill and the Hoodoos, was favorably located for his packing operation. From there, he forwarded miners' supplies from both Walla Walla and Lewiston.[29]

The transportation conditions of this period cannot be overlooked. Even though Long Jim's pack string brought necessary food and tools, which allowed the miners to continue their work, roads were necessary for the development of an agricultural economy from growing pioneer settlements. To achieve this, a territorial road was declared for the route from Walla Walla to Colville by way of the Snake River crossing at Penewawa Creek, and Colfax was to be a shipping point. The reality was that transportation remained primitive. Daniel Notman, for instance, took a week-long trip, probably once a year, to mill grain at Waitsburg. From there, he traveled to Walla Walla to purchase groceries and dry goods. Also, Palouse farmers were limited to infrequent trips to Walla Walla because the steep grade into Lewiston was impassable in the fall when harvests were completed and there was time to go for supplies.[30]

The farmer's transportation problem improved somewhat after 1874 when Henry Harmon Spaulding, Jr. built a ferry crossing on the Snake River at Almota. At this point, the river extends to its farthest point north, and the grades down its canyon walls are more gentle than anyplace above Riparia. Almota became a shipping point for the Oregon Steam Navigation Company and was the first stable outlet for agricultural produce in the Palouse. The townsite was laid out in 1877 with a warehouse and river ferry and located on the territorial road from Walla Walla through Dayton and Pomeroy to Colfax.[31]

The Hoodoo Mining District was recognized early as a legal element of this growing network. On April 8, 1875, the Nez Perce County commissioners in Lewiston heard and granted a petition by Frank Points, H.M. Dufeild [sic], and

Another symbol of intensified activity in the Hoodoo region was the construction of Woodfell School (District 61). (Latah County Historical Society, number 25-6-27)

P.W. McCabe that the trail from Camas Creek below Gold Hill to Hoodoo Camp be declared a county road. The weekly pony post from Lewiston was extended to Farmington that year, and shortly thereafter regular stage routes were organized. The drivers on those routes—Joseph Cox, Tom LaDow, Felix Warren, and Major Wimpsey—included on their schedules the mining routes to the east and northeast. This may have been in response to the miners' petitions.[32]

These regular commercial contacts suggest that an organized community existed in the upper Palouse Valley by the mid-1870s. While the precise location of "Hoodoo Camp" has yet to be documented, it is likely that the community was near what became known as "Grizzle Camp," the most notable mining boom settlement in the region. Grizzle Camp was situated on a knoll in present-day Laird Park at the end of the Palouse River bottomland. It became important as the major terminus for stagecoaches and pack trains. In earlier Hoodoo history, camps had simply developed adjacent to the most heavily worked claims.

Grizzle Camp took its name from Griswold's Meadow and "squaw man" John Griswold, who lived with his Nez Perce wife and their children on the ridge above the junction of the Palouse River and Strychnine Creek. Griswold had orignally settled near Viola, but the arrival of farmers forced him "farther back" into the hill country. He left the upper Palouse for good when

Much of the mining that took place in Latah County was conducted using placer equipment. Here, an unidentified miner is utilizing some of the most basic tools to extract gold: a hand-powered rocker, a gold pan, and a cooking pot to supply the necessary flow of water. (Latah County Historical Society, number 25-4-10)

the miners arrived. According to one recollection:

> Many years ago, when the first miners came through the country, there were lots of meadows, and fishing was wonderful. But then, after they'd muddied the streams and taken all the game that they could—now there weren't just a few miners, there were hordes of them—the country wasn't the same. Old Griswold, he'd shake his head and he'd say, "No good, no good." And he'd go on. He got enough skins from his beaver trapping that he done pretty well anyway.[33]

Various individuals took part in activity at Grizzle Camp, especially local farmers with produce to sell. One of them, Ed Graham, is credited with having built an eating house, blacksmith shop, and saloon as early as 1874. Graham also ran a pack train and pastured his animals in the meadows, along with stock belonging to Jap and Green Chambers who ran the stage line through Viola to the Hoodoo District. Graham's own homesteads were on the Palouse River and Meadow Creek.[34]

The St. Elmo Hotel became a prominent part of the Palouse business establishment serving as a place where mining promoters, such as Paul Bockmier, brought mining speculators to discuss investing in the Hoodoo mines. Built in 1888, the hotel boasted fifty rooms, "sample rooms" for salesmen, an elevator, and a bar. (Whitman County Historical Society)

With the exception of a hiatus during the Nez Perce War of 1877, the population in the Palouse country grew rapidly during the 1870s and 1880s. Worley, Farnsworth and Company built a sawmill at Palouse City in 1875 and started the lumbering boom in that community. The town itself was laid out by W.P. Breeding, who built a flour mill on the site in 1874. Other businesses were established when people with a variety of goods arrived by wagon and stage. However, the great impetus for growth came in 1877 when several families came to Palouse City in eighteen wagons. They had waited in Dayton throughout the Nez Perce War until an army escort could accompany them northward.[35]

During this period Chinese immigrants became a prominent feature in the life of the region. It was common to see Chinese in Palouse City, and they owned at least one laundry service and probably two restaurants. It may have been the Chinese community's connection with the Hoodoo mines that provided a ready commercial climate for the town.[36] At the mines, whites were aware of the wealth being extracted by Chinese miners on Poorman Creek, Excavation and White Pine Gulches, and China Hill. As in other Idaho gold districts, the Chinese were organized in "tongs" for their massive effort to conduct large-scale placer mining. An anecdote involving a Chinese gentleman on the Chambers Brothers stage line, reveals the extent of this mining activity:

> Then on up to [Grizzly] Camp, well, that's where they stopped the night. And they got there, Green said, "You go on in the hotel." It was quite a nice big log cabin, and it had a restaurant there too, besides the room where people could stay. . . . The next day they took him up to Green's, the end of the stage line, where he found his Chinese coolies there working the mines up on China

> Hill. . . . When they come back—the Chinaman always collected every bit of their gold dust, every time he came in, and he came every month, regular. And then he'd take the gold back down to San Francisco.[37]

Chinese mining enterprise did not end with the purchase of goods and services from Palouse City merchants. Legal rights to profits were negotiated between tong leaders and town financiers and real estate brokers. Many of these transactions were handled by A.A. Kincaid, J.G. Powers and J.H. Wiley. These three men controlled much of the area's available capital through Powers's Security State Bank established in 1878, Kincaid's Northwestern Pacific Mortgage Company, and the Palouse Mercantile. Examples of this exchange were two ninety-nine year leases taken by Charley ("Charles") Yet & Company in September 1886 and May 1887. For the consideration of $500 the company leased forty acres of placer ground, "including ditches, etc, for mining." Other similar contracts were negotiated between 1886 and 1888.[38]

It was also common for whites to hire Chinese miners to work claims. Adam Carrico employed the largest crews in the area, reportedly between 80 and 200 Chinese, to work his Gold Hill enterprises. In 1893 the *Palouse News* described whites hiring Chinese miners for their operations in the Hoodoos: "Several large parties have passed through the past week to the Hoo Doo mines, including Mr. Sam Marten of Farmington and Harry Britten with his [corps] of Chinamen." Another report, toward the end of the 1893 season stated that the Chinese paid a monthly rental fee of $75 for their diggings.[39]

Fact—or more likely rumor—regarding the richness of the Chinese diggings, the increasing nationwide economic depression in the 1890s, and the large population of whites prospecting the Hoodoos may have all combined to force the Chinese from the boom areas. Regional newspapers carried "boilerplate" stories of anti-Chinese violence in San Francisco, Los Angeles, and Seattle, especially during the congressional debate over the various acts and treaties aimed at limiting Chinese immigration in the 1880s. Major local concern was also voiced over the alleged smuggling of Chinese into the United States by way of Canada and the Colville Reservation. The concern was over the effect these illegal immigrants would have upon the labor market. Under the headline "Shake the Chinese," an article printed in the *Palouse Republican* declared:

> Business is dull, and many white people have no work, yet the Chinese laundries are busy. There is a good white laundry in town which deserves liberal patronage, because only first class work is done. People should remember Mrs. Bests when in need of laundry work, and if anyone is idle and has to go hungry let it be a Chinaman.

In its "random notes" section, another Palouse paper reported, "As a herder of Chinamen, constable R. M. Callison made a decided hit."[40]

Reports of depredations against Chinese claims began as early as 1881, and the Chinese presence in the Hoodoo District was curtailed by the mid-1890s. Although there were many abuses of Chinese by whites, only one prosecution for such violence is known to have taken place in Nez Perce County, which then included present-day Latah County. It involved the murder of three Chinese vegetable peddlers by a white man. In the fall of 1884, their bodies were found twelve miles east of Palouse. Ab Galloway immediately fled the area but was later arrested at Mount Idaho in Idaho Territory, wearing a pair of rubber boots missing from the Chinese camp. Other actions further incriminated him. He had, for

example, "passed by his usual trading place, Palouse City, and gone 12 miles away to Farmington to purchase his supplies which he paid for in gold dust." Galloway was arrested and spent a total of 189 days in jail. He awaited trial for 180 days because there were only two district court sessions per year. At the trial, Galloway was acquitted. Upon his release he supposedly said to the undersheriff, "Could they do anything to me now if I told you who helped me do the job?" The episode was typical of events that drove Chinese miners from diggings in gold fields throughout the Pacific Northwest.[41].

The roles of the earliest prospectors, packers, tradesmen, and Chinese in the Hoodoos can only be interpreted through circumstantial evidence because precise documentation about early placer claims does not exist. Although provisions for establishing legal rights to placer ground in the public domain were established in 1870, the extant record of prospecting in the Hoodoos is as shadowy as the miners themselves. It was not until men with substantial investments undertook the risks of concerted mining development that the documentary evidence about individual claims begins.

The first registered claim in the Hoodoo Mining District belonged to an association that called itself the Palouse Mining Company. In 1884, eight men located the maximum allowable placer claim of 160 acres between the forks of the Palouse River and Sowbelly Gulch and had its boundaries laid out by the county surveyor. The association was composed of farmers and miners of the upper Palouse Valley and businessmen from Palouse City.[42]

This activity marked the beginning of the first documented rush into the Hoodoo Mining District, although as many as 300 Chinese may have already been at work there. Just what sparked the rush is not known, but between August 1884 and July 1885, ninety claims were registered. Of these, sixteen were made by associations of two or more miners, including that of C.H. Clark and the Carrico Brothers from Gold Hill. Many of the individual claims were also located by Palouse City businessmen and upper Palouse Valley homesteaders.[43] Although the locations of many of the mined drainages have been forgotten, the rush seems to have been to the well-known sites between Poorman Creek and the North Fork Palouse River above White Pine Gulch.[44]

The progress of the rush was closely followed by the newly established *Palouse News*, especially since its managing editor and publisher, S.G. McMillin, was the district recorder, and the standard location notices for the miners were probably printed on the *News* press. The following items were scattered throughout the paper's June 5, 1884 issue:

> More confidence is being expressed everyday of the quality of the quartz ledges of the Upper Palouse, by men who are now opening up the ledges that have already been discovered; and quite a company are now in the area [?] prospecting.
>
> Several years ago $16,000 was taken out of a placer claim a little below the quartz ledges. Old and experienced miners say that the indications are better than they have seen at the Coeur d'Alene mines and that it will be but a short time when there will be a good camp in that section.
>
> The Hoodoo Mines are creating more excitement every week. Everything looks favorable for a general "boom" in a short time.
>
> W. S. Reider has been up the river prospecting.
> Everyday the stage from Moscow comes loaded with passengers bound for the mines.[45]

The Hoodoo Post Office, formally established in 1890, had its name changed to Woodfell in 1903. Constructed by John Jacob Johnson about 1900, this building intermittently served miners in the Hoodoo District until it closed in 1910. Here, the building is pictured as it appeared in the 1960s. (Soil Conservation Service)

The claim notices themselves reveal the extent to which this area had already been exploited. The Palouse Mining Company claims were mapped by reference to the "Hoodoo and Sowbelly cabins." Their claim to water rights stated: "Commencing at this stake and running along the line of the old Hoodoo Ditch as far as it runs. Thence along the mountain side to Poorman Creek. We claim 2500 . . . of water for mining purposes." Other notices also recognized the advantages of reclaiming old ditch systems by claiming water rights above gulch crossings.[46]

The flush of excitement aroused in the summer of 1884 carried throughout the winter. The snowfall that year seems to have made little difference, as prospecting continued heavy through the month of December. Even in January and February 1885, prospectors discovered nine new claims. Of these, one was located by Oliver Hazard Perry Beagle, the developer of Beagle's addition in Moscow. It is notable that several winter prospecting ventures took place on Grizzle Bar and close to Grizzle cabin, all comfortably near the amenities offered at Grizzle Camp.[47]

General interest in working the Hoodoo placers continued throughout the following two seasons as well, but with diminished enthusiasm. In 1886, twenty claims were located, primarily during September and October. In 1887, miners located only seventeen new claims. Most of these were speculations by well-known prospectors and mining men who hoped to sell them to Chinese.[48]

Soon after the fire of 1888 businessmen began to rebuild downtown Palouse, this time in brick and stone. The Swarts House, later the Commercial Hotel, proved to be one of the most significant of the new buildings. It became a location where entrepreneurs struck deals for mining enterprises in the Hoodoos. (Whitman County Historical Society)

In the fall of 1886, W. G. Connor began his thirty-year association with the Hoodoo Mining District and served much of this time as Deputy Mining Recorder. Connor had actually begun registering mining claims in the fall of 1885 when he served as recorder on Last Chance Creek. From then on, his filings at the county courthouse accurately reflected the information he first recorded in the *Hoodoo Mining Record,* a set of books that were maintained in the mining district.[49]

General population and economic growth in the Palouse River Valley were rapid during the 1880s. With a population of about 200 in 1882, Palouse City literally moved its business district from the hill above the river to the bottomland along the banks. The thriving grain mills were driven by waterpower, but it was the use of the river in sawmilling which made the lumber industry most important. The thickly forested slopes near the Palouse River provided timber for great log drives downstream as far as Colfax, The construction of the Northern Pacific Railroad to Palouse City in 1888 provided long-awaited economical access to markets. By 1891 the town's population was about 1,200, and the three sawmills employed 200 men. Agricultural production benefited from rail transport as well. Prior to rail shipping, agricultural products were wagon-freighted 27 miles to the Almota bar to

await river transport. Slush ice in the Snake River often caused delays, and the two portages in the Columbia River required further transshipment.[50] With a booming lumber industry and reduced costs for railroad transportation, Palouse City became one of the most important towns in the Washington Territory.

Without exception, the entrepreneurs and commercial operators who survived the instabilities of the pioneer period held interests in Hoodoo placer mines. Those men already mentioned who were among the earliest arrivals – Kincaid, Powers, Wiley, Farnsworth, Preffer, Northrup, and F.M. Smith – were able to expand their control of the town's capital economy by diversified holdings and interlocking directorships in real estate, insurance, banking, mercantile, and livery and transport. Later arrivals who joined the business community also fitted this pattern. J.K. McCormack, for instance, was a partner in a claim "on hill between Hoodoo Gulch and Greenhorn about one-half mile from Palouse River." He was also cashier at the Security State Bank and an insurance agent representing companies headquartered in San Francisco, London, Philadelphia, Hamburg, Hartford, and Portland. Adolphus Galland, owner of Galland Brothers grain dealers, and the Galland Trading Company, held a claim on the North Fork Palouse River.[51]

Palouse City businessmen could exploit the wealth of the Hoodoo Mining District because they had intimate contact with the prospectors and miners themselves. The values of placer claims were highly guarded secrets, and the development of mines required discreet marshalling of labor and money. Since there were no minerals exchanges, where anonymous investors could take stock in bonded enterprises, interested parties needed a place to conduct business. The Swarts House catered to miners and mining speculators, and it is likely that mining transactions were completed there. Palouse City probably remained the focus for gold mined from the Hoodoos because the Swarts House provided a convenient meeting place during the pioneer period.

The Swarts House was a brick hotel with 50 rooms. It was built after the destructive 1888 fire in Palouse City and located on Main Street at the east end of town. Its owner, A.J. Swarts, originally homesteaded near the mouth of Meadow Creek in the upper Palouse River Valley, but he moved to Palouse City and ran the hotel until 1894. Thereafter, it was known as the Northern Hotel.

The importance of the Swarts House in the history of Hoodoo mining development can be seen in an account of men who lived there or ran businesses nearby. In 1891 for example, several of the boarders had registered Hoodoo Mining District claims, and they list themselves in a wide variety of occupations. John English, laborer; Emery S. Brents, lumberman; John Butzow, carpenter; and F.M. Smith, bookkeeper, would have had contact with Harry Rice, part owner in the Ihrig and Rice butcher shop; George H. Sheldon, foreman of the *Palouse News;* and miners George W. Speake and William King. D.C. "Dud" Tribble was in a furniture store partnership called "Lamb and Tribble" with J.C. Lamb, but he was also a well-known prospector in the upper Palouse River Valley along Rock Creek, Hatter Creek, and Turnbow Gulch. Another boarder was W.F. Chalenor, an active mining promoter as well as manager of the C. and C. Milling Company, located near the

In 1940, Northwest Goldfields, Inc., of Spokane, Washington began mining operations on the Palouse River with a four-and-one-half foot bucket dredge. In 1942 the War Production Board confiscated the dredge's two 300 horsepower diesel engines for America's effort in World War II and work ceased. Briefly, the dredge was refitted in 1947 and used for a few months around White Pine Gulch and Poorman Creek before being abandoned. (Latah County Historical Society, number 25-4-13)

Northern Pacific Railroad depot. Though not a Swarts House resident, J.C. Northrup housed his business, Skeels and Northrup, just opposite the hotel on Main Street. He was a long-time prospector who held extensive mining interests in the Hoodoo and Gold Hill districts in addition to his Palouse City businesses that involved blacksmiths, machinists, wagonmakers, and liverymen.[52]

At the Swarts House, speculators and prospectors had regular opportunities to create investment schemes. This was particularly important for the formation of mining associations, which were the basic organizations for placer claim development. Few of the boarders listed themselves as miners, and the variety of men's occupations shows that many of those willing to labor over a sluice were placed in close contact with those willing to provide a grubstake. Thus, labor as well as capital were available at the Swarts House.

As a part of its move to create a timber monopoly in the region, the Potlatch Lumber Company purchased the mills at Palouse. This view of the facilities was photographed about 1904. Subsequently, this operation was closed when the company opened its massive new mill at Potlatch, Idaho. (Whitman County Historical Society)

General enthusiasm for placer mining in the Hoodoo District diminished after 1887, but new settlements at Freeze, Starner[53], Cove, Chambers Flat, and Woodfell assured the continued interest of Palouse City businessmen in the gold potential of the upper Palouse Valley. E. H. Orcutt, whose *Boomerang* newspaper closed in 1889, reported that as postmaster he expected to start a new mail route up the river in 1890. Further interest was sparked when a Northern Pacific spokesman announced the possibility of a spur line from Palouse through Starner and across the divide to the new Ruby Creek mines. This rumor prompted a call for even more minerals development. An irregular letter, or column, in the *Palouse News* called "Hoodoo Nuggets," transmitted tidings of progress and described conditions at the mines. Though these letters were signed simply "Hoodoo," they may have been written by someone like long-time miner Frank Points, once described as a "cultured Virginia Gentleman." Hoodoo Nuggets contained a homely blend of anecdotes, weather reports, promotions, and information on placer technology.[54]

With the coming of the railroads and the dramatic influx of settlers in the 1880s, agriculture and lumbering were the most important economic assets in Palouse City and the upper Palouse River region. Nevertheless, boosters of the region's economy continued to tout mining as the most important of "other resources." One brochure claimed Palouse City was the

market for about $60,000 in gold dust each year.[55]

New strikes in 1893 caused another rush for Hoodoo gold. After patient labor all spring and summer on a claim staked the previous fall, John K. Truax "made a clean-up of nearly $100 off a piece of ground 9 by 16 feet." His claim was on the North Fork about a mile above White Pine Gulch. While many were looking for the source of Hoodoo gold, Truax said he was merely "looking for the place where some of it stopped in paying quantities." Newspapers reported intensive mining activity all over Latah County with "unprecedented numbers" of men looking for ledges and sending out samples for assay. One story reported that Dud Tribble's strike at the head of Meadow Creek caused the gold fever. This rush to the Hoodoos, however, was undoubtedly spurred by the nationwide economic depression that hit the region around Palouse City in the wake of the Panic of 1893. The magnitude of the disaster was underscored by the collapse of the McConnell-Maguire Company in Moscow, the region's largest general mercantile house. When that year's harvests were destroyed by unseasonable floods, it was certain that larger numbers of men would be seeking the $2.00 per day they might earn by working in the placer mines.[56]

Perhaps for the first time in the Hoodoo Mining District, the 1893 rush attracted an extensive speculative interest by a company that wanted to import machinery to intensively mine a considerable distance along the deep river gravels. Previous associations of businessmen, miners, and investors had developed one or two claims of which none exceeded 160 acres. But this new conglomerate had a considerably larger purpose. Although little is known of the details, the venture was underwritten by financiers in Minneapolis and Spokane; locally, the organization was represented by W.H. and J.N. Muncy (or "Muncey") from Tekoa, Washington. They located and bonded about four miles of the North Fork and Palouse Rivers as far down as Bluejacket Creek. They apparently planned to employ seven men to operate a dredge to mine the river bottom. By this time, it was clear to Hoodoo miners that any profitable mining operation would require working the gravels to a depth of at least eighteen feet and constructing a five-mile-long access road from Grizzle Camp. This new venture was an ambitious undertaking with the new company planning to invest $26,000 in machinery alone. As a hedge against failure at a single location, they also acquired rights to mine on Columbia and Snake River bars. Locally, however, their plans failed. By 1898 the required road up the Palouse River had only reached Strychnine Divide.[57]

It was more common for homestead families to take up their picks and shovels after the fall harvest and head for the Hoodoos. This was especially true for settlers in the upper Palouse Valley east of Deep Creek. For them, the only real cash crop was timothy hay, and their homesteads first had to be cleared of heavy timber. Therefore, any money that the family might earn while panning gold was important. Many of these would-be settlers and gold miners simply did not "make it" unless they were involved in a variety of activities to support their homesteads.[58]

Though the numbers of annually registered claims diminished somewhat after the excitement of 1893, they remained consistently high through

1897. Miners continued to work the heart of the district, although most of their efforts were concentrated along Strychnine Creek and its tributaries, as well as the virgin ground above White Pine Gulch on the North Fork of the Palouse River. The ditch system of White Pine Gulch, known as the "Old China Ditch," was reopened to mine China Hill for about one-half mile along the rim of the west slope. John English, E.K. Parker, H.G. Rice, Truax and the Taylor brothers prospected from Moscow Gulch northward, while "old-timers" Pat Flynn, E.S. Brents, C.W. Sanderson, and Frank Taylor joined miners in the Strychnine area.[59]

In June 1898 John and Charles Taylor located their Mother Lode and Mother Lode Extension claims. Ultimately called the Mountain Gulch Group, these claims became the heart of the only extensive gold lode development in the Hoodoo District. By the middle of the 1899 season the Taylors' activity caused another rush to the Hoodoos. New claims were made on land surrounding the Mother Lode properties, up and down Mountain Gulch, in Moscow and Eldorado Gulches, and northward to the prominence that became known as Gold Hill. This flurry of activity was not restricted to the upper North Fork region. Many fortune-seekers with years of experience in the Hoodoos continued to locate claims in the more traditional areas.[60]

In an effort to open new ground in the upper part of Poorman Creek, A.J. Choat of Palouse City, along with E.L. Hemingway and his son Bertram, located on Rocker and Hemingway Gulches. The Hemingways make a good case study of the way in which "gold fever" affected men on the American frontier. Born in New York, E.L. Hemingway headed west at the age of fourteen. At The Dalles, Oregon he worked on the toll road and was a merchant. Next, he raised stock along the John Day River, but in 1860 he joined the rush to the Cariboo mines of the Kootenai District. In 1878 Hemingway had a narrow escape from Indians at Cayuse Station during the Bannock War. During the following year he secured title to a Snake River bar, which became known as Hemingway's Landing, at present-day Illia. Since this river bar was at the only accessible point to the Snake River for at least twenty miles, it became a location for commercial activity. In addition to a warehouse, store, orchards, and a post office, Hemingway's Landing was a connecting point on the daily stagecoach service between Dayton and Colfax. Hemingway undoubtedly heard about mining activity throughout the region when news of fresh discoveries in the Hoodoos reached his Landing. In 1898 at the age of fifty-five, E.L. Hemingway and his son struck out for the Hoodoos to take up a claim as placer miners.[61]

In addition to the more sophisticated locally-based mining developments, minerals promoters and professional prospectors began locating tracts during the late 1890s.[62] Seasonal labor for mining enterprises was less available when industries, especially logging, became stable enough to provide regular paychecks to employees. Also, increased capital expenditure was necessary to mine beyond the shallow surface deposits. For these reasons joint stock ventures began to play a larger role in the Hoodoo Mining District. These were companies which generally sought wide-scale popular investment, and they were prepared to sell enormous numbers of shares. Typical articles of incorporation authorized their sales of 1,000,000 to 1,500,000 shares at "cookie jar"

rates (shares selling anywhere from two cents to $5.00 per share—with $1.00 per share being the most common price).[63]

Minerals exploration firms from the Midwest or East often operated through a local agent or promoter who located, bought, and resold claims. At times, these promoters simply relocated potential ground after the original locator had failed to meet the year's assessment work requirements. They seemed to be intruders in the mining community, and so they were known in the districts as "sharpers." But miners also recognized that promoters were both employers and attractors of investments. Promoters believed in themselves, and deals were often struck on properties "without any showing whatsoever." One of the best known mining promoters was Paul Bockmier, Sr. of Palouse. Previously active on Gold Hill north of Princeton, he claimed that many good prospects went begging because "the old timers wanted too much money for them to induce capital to come in and develop them."[64]

Interest in the Hoodoo Mining District continued at various levels after the turn of the century. Established in 1905, the Potlatch Lumber Company consumed much of the excess labor that otherwise might have supported continued mineral development. The mines, then, were essentially the province of a narrow group of local mining interests and those who considered themselves miners. The most active and identifiable of these groups were the Taylors' Mountain Gulch Mining and Milling Company, J.C. Northrup and his varied concerns, and the Blue River Mining Company.[65] Upon being organized in 1905, the managers of the Blue River Company declared their grandiose purpose:

> [to] buy, sell, lease, operate and develope [sic] mining properties, . . . real estate, water rights . . . and to locate each and all . . . to buy, sell, lease, operate and develope smelting, concentrating, and milling plants, railways, and steamers and steamship lines; to buy and sell ores . . . to build, buy, sell, lease, operate and develop wagon roads and tramways; to erect and lease buildings, and to do and carry on a general merchandising business, mining, milling and sawmilling business and boarding and lodging house business. . . .[and to] be active in Washington, Oregon, Idaho, Montana, and other states and British Columbia.[66]

Such all-inclusive statements may well have been necessary to carry on minerals development in the region at the beginning of the twentieth century. Between 1900 and 1903 property boundaries claimed by Northrup and the Taylors led to a series of registration battles over the Mountain Gulch claims. In January 1904 the Taylors refiled their locations and emphasized that their company had been incorporated in 1900. Northrup, however, declared that he had amended his location notices and had actually filed first. Further, he argued that the Taylor brothers had failed to adhere to the "prudent man rule;" they had not, in other words, continued to do annual assessment work. The issue was resolved by a survey plat of the Mountain Gulch Group. However, by the time the survey was completed, Northrup had likely lost interest in the disputed property. By then, he was focusing attention on his Mizpah claims.[67]

J.C. Northrup claimed to have relocated the prominent Mizpah Creek copper mines in about 1900 and organized the Merger Mining Company to develop them.[68] In late 1905 and early 1906, the initial prospects were perfected, and Northrup began to develop his properties with the highest hopes for the Hecla and Chancellor

ledges. Northrup's prominent business partners included W.F. Chalenor, C.E. Frederick, W.M. McCroskey, W.R. Belvail, and George N. Lamphere.[69]

Work in the Mizpah mines did not progress rapidly. It took at least two financial reorganizations and a decade of work before resident managers J.C. and E.R. Northrup were able to produce the first commercial ore in 1916. The ore had to be hauled overland by wagon to the railroad depot at Harvard, Idaho. At the site itself, the Northrups had invested considerable capital to create a commercially viable enterprise. They cut five tunnels, totaling 1,550 feet in length with 375 feet of drifts. They purchased a sixty-horsepower gasoline engine, a Fairbanks-Morse air compressor, machine drills, an electric light system, tools, ore cars, steel rails, air pipes, and constructed several buildings. Development continued through 1919, though 1918 proved to be the mine's most productive year when about 79,000 pounds of copper was smelted from Mizpah ore.

Unfortunately for the entrepreneurs, the financial difficulties of creating an economically viable large-scale mining operation in the Hoodoos proved impossible. To overcome the high freight costs associated with transporting unprocessed ore, they hauled lumber over winter snow to build a mill in 1920. Nevertheless, production was halted. Miners produced copper ore from the mine in only three other years – 1924, 1925, and 1929. Subsequently, the Mizpah claims were held by the Columbia Mines Company of Spokane.[70]

The Northrups were not alone in attempting hard rock mining in the Hoodoo District. Locals like C.W. Sanderson; T.P. Jones, Potlatch Lumber Company's woods superintendent; Hugh Henry, Deary Townsite Company manager; and Ray Palmer started the Latah Copper Mining Company Ltd., of Bovill. They hired five or six men, drove three tunnels, built a cabin, blacksmith shop and powerhouse, and acquired a hoist boiler, two Burleigh drills, and a small compressor. For their efforts, they shipped out one railway carload of hand-picked ore which was hauled from the mine to the depot in a Model T Ford truck that could carry one ton per trip. Subsequently, smelter operators informed them that the ore ran only five percent per ton and was therefore, not worth handling. The Latah Copper Mining Company made no money.[71]

Development continued in other portions of the Mountain Gulch area, and claim notices were posted, especially on Gold Hill above the North Fork of the Palouse River, on behalf of the Star Crescent Mining Company, the Western Mining Company, and the Progressive Mining Company; the latter was registered in Multnomah County, Oregon and represented locally by J.D. Connor. The Taylors' mine was reported as showing deposits with ore that produced gold at the rate of fifteen to twenty dollars per ton. The mine had 600 feet of tunnels and 125 feet of incline shafts, and was equipped with a card table, boiler, engine, and a two-stamp mill which may have been replaced after 1908 by a Huntington roller mill and a Blacke crusher. The smaller independent enterprises, fewer in number than during previous years, continued to merge their interests for communal work on diggings.[72]

After 1912, the amount of placer gold produced in Latah County, with the Hoodoo District being the most significant mining region, fell considerably. Except for the years 1922 and 1923, production remained negligible until the early 1930s. By then, the Great Depression was so

Paul Bockmier, Senior, a businessman, town booster, and mining promoter, worked tirelessly for much of his life to bring large-scale, commercial mining to the Hoodoos.
(Whitman County Historical Society, Paul Bockmier Collection)

widely felt that armies of unemployed again turned to placer mining in the Hoodoo District. To resident miners in the districts, these newcomers were known as "Depression miners" or "overnighters." This time, however, there was no lucky strike followed by a romantic rush to the hills. As an attempt to help put men back to work, the Northwest Mining Association held a placer mining school at the headwaters of the Palouse River in 1932. About 2000 people attended and heard W.W. Staley, of the University of Idaho and Idaho Bureau of Mines and Geology, and Paul Bockmier, Palouse mining promoter, deliver speeches. The dean of the Idaho School of Mines, J.W. Finch, stated dismally that only one in ten prospectors would earn more than wages, and only one in a hundred would find commercially exploitable ground.

One obvious sign of prosperity in the town of Palouse was the presence of the Sever Jewelry Store. (Whitman County Historical Society)

Nonetheless, the Northwest Mining Association, based on the success of this first effort, held several more placering schools around Washington State and estimated that 5000 people attended. Further, Finch believed that 3000 men had gone to the hills seeking hold. Staley's manual, *Elementary Methods of Placer Mining,* was reprinted in five editions between May 1931 and June 1932. In the Hoodoo District, the White Pine, Moscow, and Eldorado Gulch areas were most frequently registered as placer locations. Despite the influx of miners however, the gold produced in the Hoodoo District between 1933 and 1938 only matched the amounts produced in the years of decline between 1906 and 1912.[73]

The last major effort and the largest single attempt to extract gold from the alluvial deposits in the Hoodoo District was a massive river-dredging operation undertaken by Northwest Goldfields, Inc., between 1939 and 1942. About five miles of river bed from the mouth of Moscow Gulch on the North Fork of the Palouse River to a point just above the mouth of Bluejacket Creek on the main Palouse River, were ultimately turned over by the dredge. Outlining these operations, the following summary ap-

peared in a 1957 survey of mineral resources in Latah County:

> The Northwest Goldfields, Inc., of which Mr. [L. J.] Burrows [of Spokane] was president, erected a four and one-half foot bucket dredge on the North Fork of the Palouse River in the early part of 1940. The dredge was set up near the mouth of White Pine Creek in the upper part of the North Fork. A considerable amount of placer mining at this location had left a depression which was easily converted into a pond for flotation. Dredging started June 1, 1940 and was almost without interruption until October 15, 1942, when Limitation Order I-208 issued by the War Production Board closed down the operation for the duration of the war. During this period of operation, the dredge excavated the North Fork river bed from White Pine Creek to the confluence of the North Fork and the Palouse, a distance of approximately two and one-quarter miles, and moved a total of 2,900,000 cubic yards of material. The gravel ranged from 10 feet to 25 feet in depth with an average depth of about 18 feet. Bedrock, except for local spots, was decomposed enough to allow digging of the upper two feet or so with the bucket line. An average of 18 inches of bedrock was removed in the dredging operation.

The gold content of the gravel varied particularly in areas where the dredge re-worked old placer tailings; however, the average value was about $0.18 a cubic yard. Operation costs averaged $0.07224 a cubic yard before depletion and depreciation; labor accounted for $0.03245 of this amount and material for $0.03979. Clearing cost about $0.01 a cubic yard. These averages would have to be revised upward in estimating present day costs.[74]

The ordeal experienced by the men hired to bring in the Northwest Goldfields dredge illustrates the extent to which conditions in the Hoodoos forestalled the introduction of large-scale heavy machinery. During the winter of 1939-1940 equipment was trucked piece by piece over the muddy Palouse River road. The heavily-loaded vehicles mired to their flatbeds in swamps; crews alternately roasted and thawed around the smoky campfires they lit each night. Frank Milbert, who was also hired to maintain the dredge during operations, contracted to haul the equipment on A-6 International trucks connected by cable to Caterpiller tractors.[75]

When the dredge began excavations the following June, it pivoted on a forty-eight foot long spud driven into the riverbed and made a cut eighty to 120 feet wide. After several passes, a path as much as 600 feet wide was cleaned. The operation employed twenty men, including miners who resided in the district and a number of timber cutters.[76]

When the War Production Board halted production of the dredge in October 1942, it also confiscated its two 300-horsepower Cummins diesel engines for the war effort. Later, these power sources were replaced by a single 600-horsepower GMC diesel engine. Briefly, for three months in 1947, the dredge was operated again. Later, the equipment was leased by Harold Behrens; he used it for about eight months in the area around White Pine Gulch and the lower portion of Poorman Creek. Since the 1940s, placer mining in the Hoodoo District has been mostly recreational and sporadic. A few old-time miners, like Pete Doffner and Bill Freeberry, continued for some time to live along the North Fork Palouse River.[77]

In summary, there were four periods of Hoodoo Mining District history during which there were significant changes in placer mining technology and economics. The first, dating from the 1860s through the 1880s, was typified by prospecting discoveries and efforts to exploit the paying ground by hand methods. The earliest

prospectors were migrant explorers, searching for gold on a north-south axis along the flanks of the Rockies. Some came from mines in California or other Pacific Northwest gold fields, and some were merchants who came west to supply miners and settlers. Early Hoodoo connections with the western frontier economy tended to be toward the south with the riverport of Lewiston at the junction of the Clearwater and Snake Rivers, the southwest to Walla Walla, and the west to the new town of Palouse.

In the second period, capitalists in Palouse City strengthened the economic connection of the Hoodoo Mining District with their town. Development of minerals resources along with lumbering and agriculture was considered vital to their community. They invested in expansions of the hydraulic network, which had been proven successful by Chinese miners. The early homesteaders, especially those of the upper Palouse River Valley, also looked to the placer mines as a source to supplement their incomes. After floods and the Panic of 1893, the mines were an essential part of economic recovery and even survival. This second period lasted from the mid-1880s until the mid-1890s, when the regional economy diversified and the labor force was redirected primarily through the establishment of the lumber industry.

The third period, from the mid-1890s to the 1920s, was typified by the attempt to establish the commercial and industrial value of the Hoodoo Mining District. Various corporate interests moved to control prospected claims and tried to introduce labor-saving machinery that could mine deep placers or hardrock veins. Local promoters were part of these efforts. The most successful of these organizations were the Mountain Gulch Mining and Milling Company managed by Charles and John Taylor of Garfield, Washington, and the Mizpah copper mine, promoted by J.C. Northrup.

The fourth period of Hoodoo mining activity was the result of massive unemployment during the Great Depression. From 1933 through 1938 renewed individual efforts at placer mining produced some gold, but these efforts were abandoned as the general national economy began to recover prior to World War II.

During each of these periods, many men considered themselves to be members of the Hoodoo Mining District and miners by occupation. They usually resided on or near their claims and spent a major portion of each year mining.[78] Some also worked mines in other local districts on Gold Hill, Ruby Creek, or Swamp Creek.[79]

Finally, the dredging of the Palouse River in the early 1940s should be viewed as a distinct event that did not typify the historical development of placer mining in the Hoodoos. Its importance, however, should not be underestimated since it provided the best evidence for the extent of gold deposits in the district. It remains today as the most visible reminder of the long history of mining activity in the Palouse River Valley.

The Technology of Placer Mining

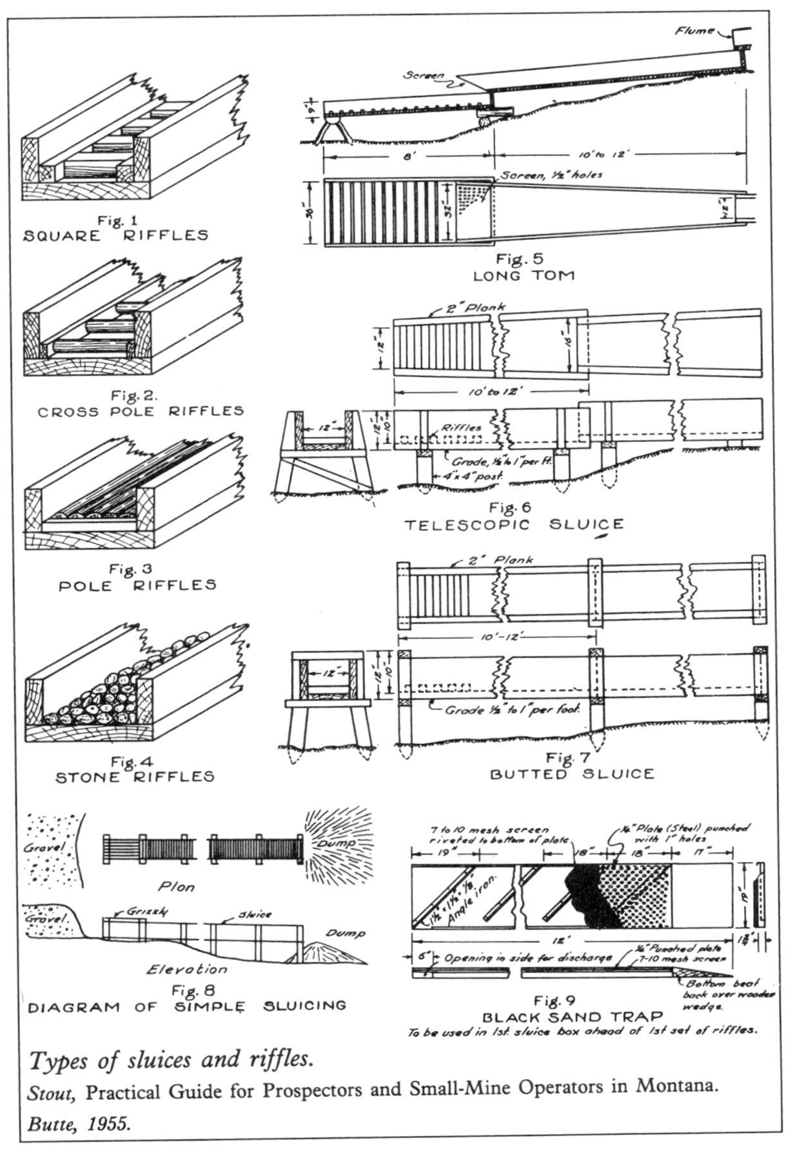

Types of sluices and riffles.
Stout, Practical Guide for Prospectors and Small-Mine Operators in Montana. *Butte, 1955.*

Placer mining is the method used to extract minerals which have been naturally removed from their original geologic matrix and have accumulated, usually in alluvial deposits. Erosion causes heavy metals to settle toward the bottom

of deposits and to build up in crevices or behind irregularities in bedrock. Gold is usually obtained by washing the gravels with water, though in some desert areas the ore was tossed into the air so that wind could winnow the metal dust. Placering methods are very old, having been used by ancient Egyptians, the Romans in Spain, and in early India.[80]

Methods employed in the gold rushes in northern Idaho were, however, based on innovations developed in California. These techniques drew upon Spanish traditions that were modified to concentrate large amounts of water on particular gold-bearing deposits. As the shallow, easier-worked surface diggings in California were exhausted, men began to explore other areas of the American mining frontier, taking with them their knowledge of placering techniques. After 1863, contemporary observers estimated that at least 20,000 men left California diggings for Nevada, Utah, and other areas in the mountain West.[81]

Many of these men, such as Elias D. Pierce, who first discovered gold in Idaho, considered themselves prospectors who located mineral deposits, as opposed to miners, who undertook development of mining enterprises. Versed in the physical characteristics of the landscape in which mineralization frequently occurred, prospectors applied a battery of tests that could indicate which minerals were present and give a rough estimate of their value. Tools carried by prospectors indicate the variety of tests they could perform. Typically, a prospector might have a shovel, quartz pick, common pick, geologist's hammer, moil (a small drill), four-pound hammer, prospect pan, drywashing sheet, sample bag, tube drill, charcoal block, blowpipe, candle, half-skillet, hornspoon, magnifying glass, acid bottle, mercury bottle, test tube, and knife gauge. With these tools, a prospector could perform a primitive assay, such as blowpipe analysis, and discover approximately how much mineral could be expected per ton of ore in a given location.[82]

Placer gold was sought because it could be mined with only "relatively little" knowledge of geology and with greater ease than gold vein deposits. In the absence of obvious indications, lode deposits could be inferred by searching for "float" particles and scales carried away from a mineralized outcrop by water action.[83] The precise form of the particles was indicative of the abrasion experienced when the particles traveled from their point of origin.[84]

Placer deposits are classified according to how gold has been transported and sorted by erosion and the form in which the deposit settles. Gold separated from the geologic matrix, but not subsequently transported, is concentrated in residual placers. Sorted and resorted deposits, like those in the Hoodoo Mining District, are called hillside, creek, gulch, and bench placers. Hillside placers are old and may be the result of geologic uplifts and redirection of streams which originally laid down the deposits. Creek placers are best known and are represented by the workings along the North Fork of the Palouse River bottomlands. These are well-sorted, with gold usually found in bedrock joints and crevices. The deposits generally run parallel with the water flow. Bench placers contain gold deposited in benches or terraces during the downward cutting of the stream. They are usually located high up from the valley floor in well-abraded gravels.[85]

A likely prospect was initially inspected by panning or dry washing. Dry washing involved plucking and puffing on a sample spread out on a sheet until only the heaviest fragments remained. Only the richest claims were worked

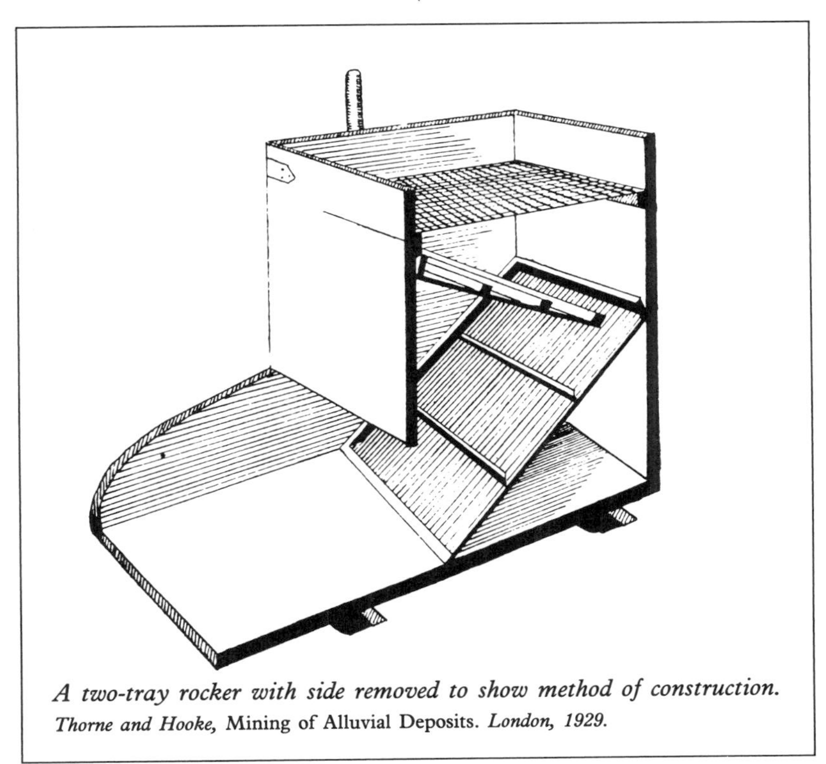

A two-tray rocker with side removed to show method of construction.
Thorne and Hooke, Mining of Alluvial Deposits. *London, 1929.*

using a pan, since estimates state that, with experience, the most gravel that a miner could hope to wash in ten hours was one cubic yard. Even in the early days in California where claims were only 100 square feet, the pan was quickly discarded in favor of more efficient means. In panning, the water was used to sort the sample. Using a sheet iron pan with sloping sides and a flat bottom, the miners swirled the material around, washing the lighter material over the edge. Since this type of pan was also used for cooking, most manuals warned against allowing grease to build up on the bottom; if this were to happen, it would make it difficult to retain fine gold in the drag. If the pan had a copper bottom, mercury could be rubbed on it, and fine gold could be retained through amalgamation.[86]

If the deposit appeared to be worth working, any of the placer methods designed to move large amounts of gravel might be employed. First, a miner might use a rocker device; this was only slightly more complicated than panning, but it allowed profitable working of low-grade gravels by an association of four men. It was generally assumed that a rocker could handle five times more gravel per day than a single man using a pan. While two men dug, the third dumped dirt and water into the rocker; the fourth man rocked the mechanism back and forth. If the soil contained clay, the clumps could be fed into a trough along with the water that ran into the rocker, breaking up the mass.[87]

A rocker was a wooden box with a handle resembling a cradle mounted on a slight incline. At the top of the rear half of the device was the

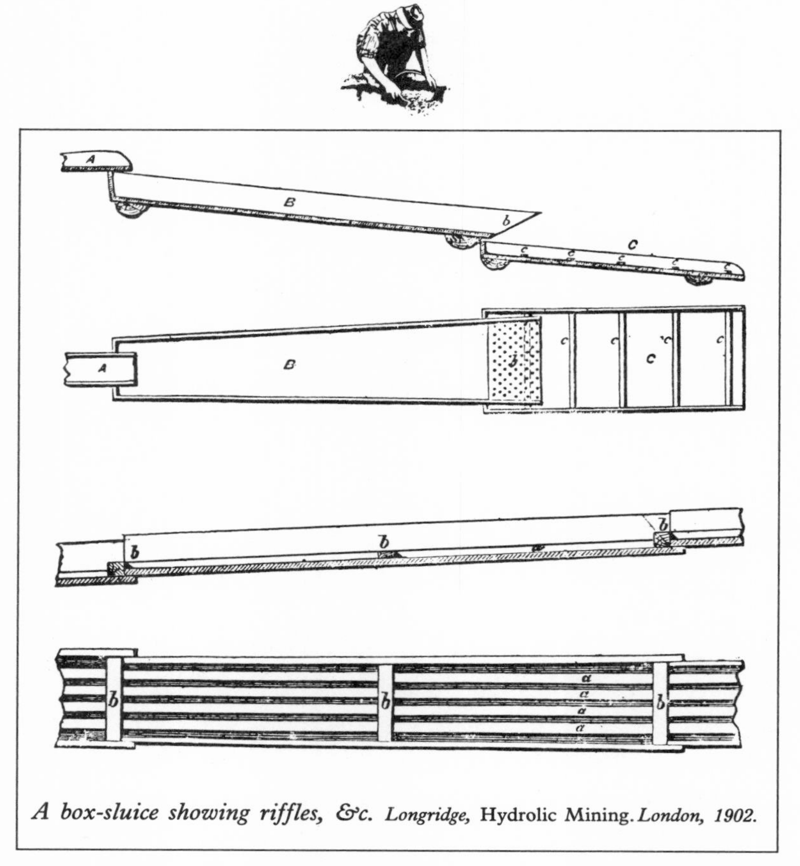

A box-sluice showing riffles, &c. Longridge, Hydrolic Mining. *London, 1902.*

hopper, a piece with a bottom of perforated sheet metal called a grizzly. Its one-half inch holes allowed the ore and water to fall through to a slanted apron, which further broke up the material and caught some gold in its sagging, canvas covering. The remainder of the material was transported to the rear of the rocker bottom. The bottom was covered with a piece of carpet or burlap. It also contained a lateral frame, usually made of square, wooden bars called riffles. A short, sharp jerking motion of the rocker and the constant flow of water would then cause the tailings material to wash over the riffles. In the process, the gold and fine sand would be caught behind the fabric carpet or burlap. During cleanup, the fabric was rinsed in a tub of water, and the gold and sands in back of the riffles was removed. The sands were further washed in a pan as concentrates. Mercury was also used to create amalgam on the bottom of the rocker. The amalgam required treatment in a retort, or else it had to be squeezed through a canvas bag.[88]

Another, more efficient and widely applied method utilized sluice boxes, or simply "sluices." Sluices functioned much like rockers, except that instead of using a rocking motion, the gravel was washed over a lengthy series of riffles, much in the manner of an artificial creekbed. Sluices were troughs, generally twelve to sixteen feet long, one to five feet wide on the bottom, and built from rough planks about one inch thick. The bottoms of these troughs held riffles made from wooden bars, wooden blocks, angle iron, or large heavy boulders, depending on the coarseness of the gravel and clays that had to be broken up. On Gold Hill, for instance, one sluice box measured three to four feet wide and made use of railroad rails and logs as riffles. Sluice boxes were built in such a way as to telescope into one another. When arranged in a series they were called a string.[89]

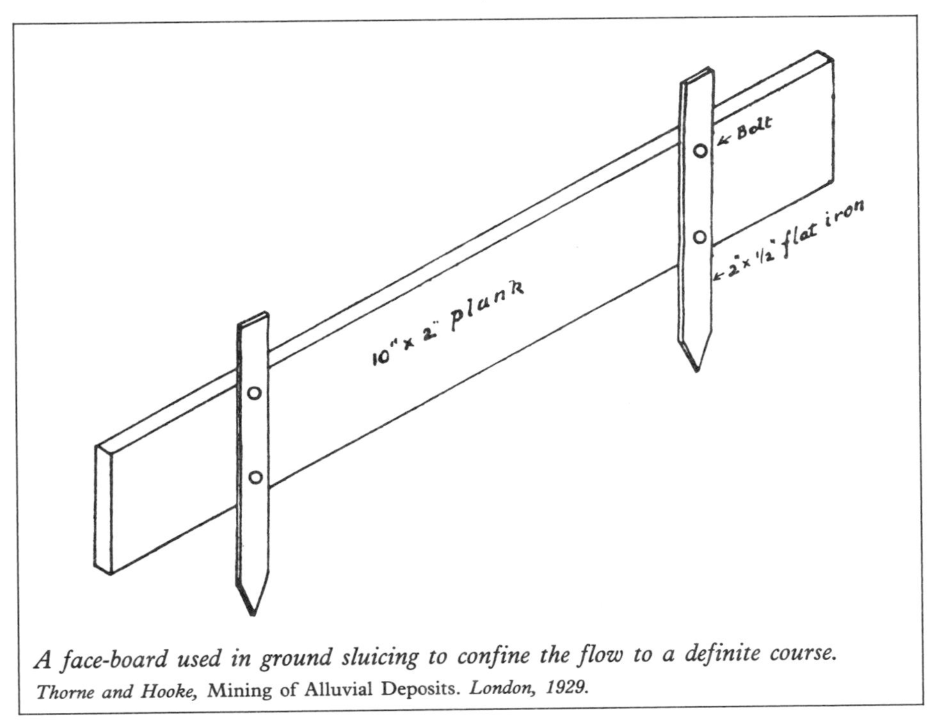

A face-board used in ground sluicing to confine the flow to a definite course.
Thorne and Hooke, Mining of Alluvial Deposits. *London, 1929.*

Several men were required to work a sluice, "some to strip the sod, some to dig and wheel, one to throw out pebbles and boulders with a sluice fork and one to throw away tailings." The importance of understanding the properties of inertia and the use of a sharp pick to undercut the gravel was essential. "A greenhorn has a short thick, stubbed pick," noted the Virginia City *Montana Post* in 1865. "He stands on top like a chicken on a grain pile; gets out one rock and finds he has another below it." Once the gravel had been dug, it was shoveled into the sluice's head box. On the bottom of the head box, the grizzly caught the coarsest gravel as the flow of water washed the finer material down the length of the sluice. The string was laid as straight as possible, and the incline was adjusted to slope from two to fifteen percent depending on the cohesiveness of the matrix and the amount of water.[90]

Sluicing, however, was far more complicated than simply erecting a system of boxes and dumping in the gravel to be washed. The most important problem was the disposal of tailings. These enormous piles of cleaned sand, silt, and gravel were known to cause stream deltas and shoalled rivers, occasionally closing claims because sluices could not be set up at a sufficiently steep angle. To prevent this, sluiced claims were worked in an upstream direction. The upstream bank receded as its earth was shoveled into the sluice. The sluice was extended to the working face by bringing forward the boxes from the discharge end. With planning, it was therefore not necessary to move the entire string, and tailings were always deposited on already worked ground.[91]

Variety in the application of sluicing methods was limited only by individual ingenuity. For example, rockers were used in combination with sluices. A more compact sluice, called a "long

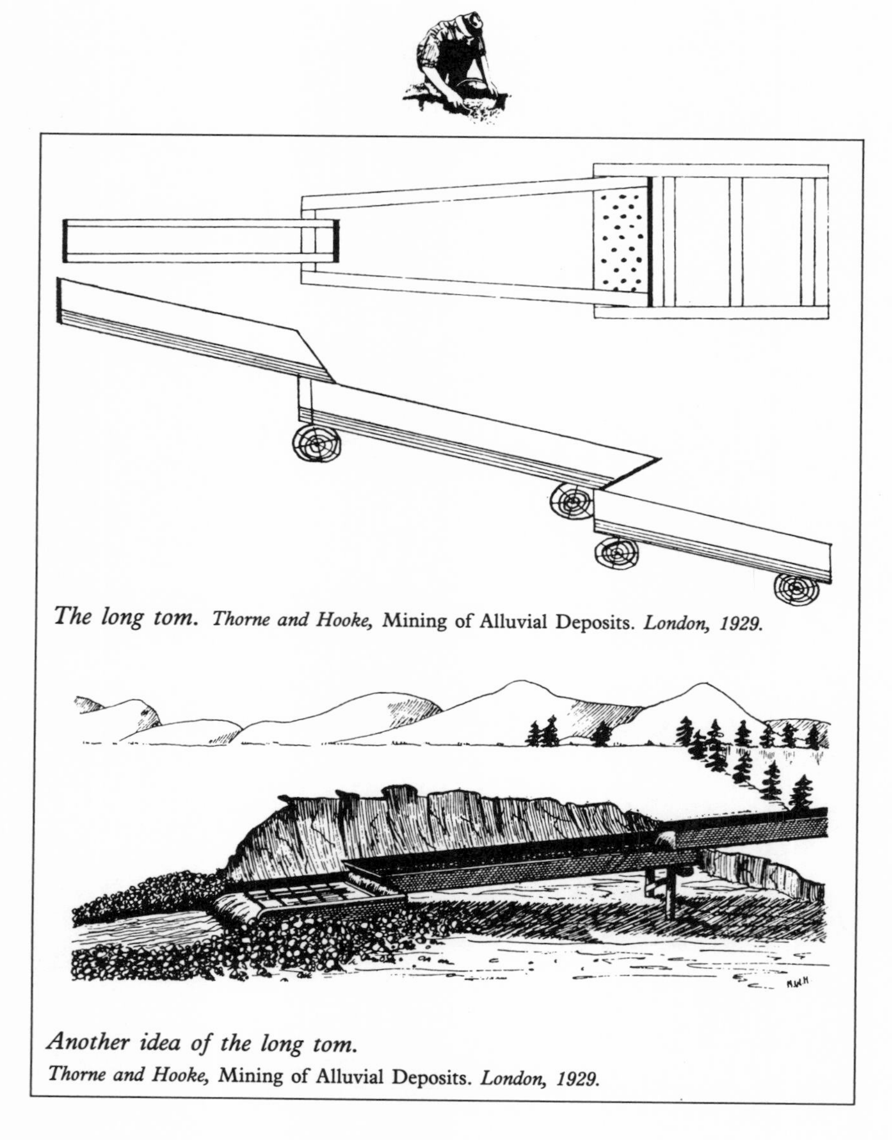

The long tom. *Thorne and Hooke,* Mining of Alluvial Deposits. *London, 1929.*

Another idea of the long tom.
Thorne and Hooke, Mining of Alluvial Deposits. *London, 1929.*

tom," used only two sections of boxes. The Risdon Iron Company of San Francisco produced angle iron riffles that were supposed to create deadwater under the angle to help save gold. Wear on sluice box bottoms could be reduced by facing the clear lumber frame with rough planks. But one of the most widely adopted innovations in large sluice systems was the use of an "undercurrent;" a wide, shallow sluice designed to wash fine material using a much decreased velocity of water. A grizzly fed the undercurrent near the end of the sluice, allowing sandy material to be washed in a thin layer over small riffles.

Numerous small manuals about placering used by even the most experienced miners were full of suggestions on how to handle large volumes of water and gravel in the rough and ready sluice box.[92]

Sluicing required an enormous volume of water brought to bear on the boxes by a system of dams and ditches. Building the appropriate canals, ditches, flumes, and impounding dams was a difficult business that required engineering skill, capital investment, and cooperation. In the Hoodoos, the richest and most accessible deposits were on hillsides, benches, and gulches along the

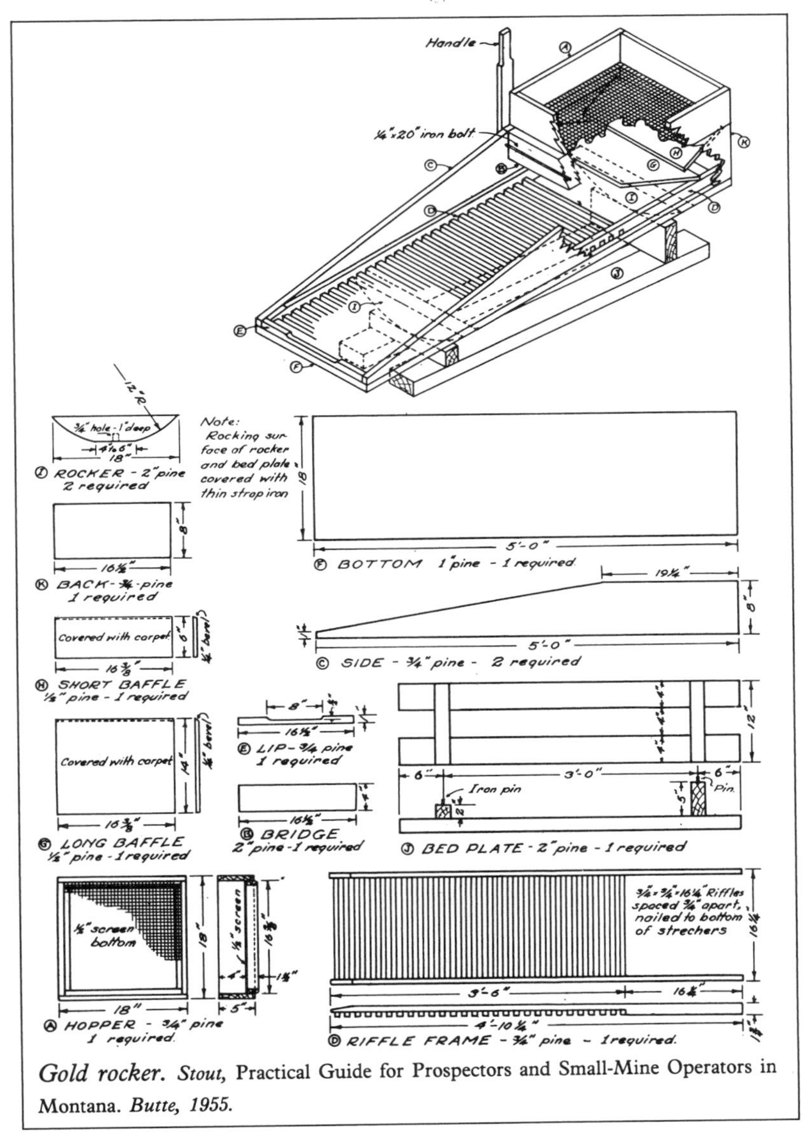

Gold rocker. Stout, Practical Guide for Prospectors and Small-Mine Operators in Montana. *Butte, 1955.*

western slope of the North Fork of the Palouse River. By and large, the drainages in that area contain some flowing water year around, but none contained enough to work placer deposits with sluices. Therefore, the earliest filing of claim notices also located water rights that specified sites for ditches, as well as the amounts of water that would be transported. Transported water was measured by "the miners' inch," defined as the volume of water flowing through a one-inch square hole in a board six inches below the stream's surface. Although the method of measurement has varied from time to time and from state to state, the current legal value in Idaho for a miners' inch is nine gallons per minute; in other words, fifty miners' inches equal a flow of one cubic foot per second.[93]

Ditches were the main structures used to col-

lect and transport water to a miner's diggings. Miners built wooden flumes in rocky terrain, or across deep canyons, but this alternative proved to be expensive and appears to have been unnecessary in the Hoodoo District. Nonetheless, the construction of ditches proved to be significant engineering feats, since they usually transported water over relatively long distances and at a slight, but steady grade at a uniform width and depth of excavation. An ideal ditch was subject to little erosion, evaporation, and seepage. Further, it had to transport a large volume of water throughout the year. It had to be located as high as possible, but below the spring snow line to prevent blockage when snows melted on exposed slopes. All the small drainages along the ditch's length were also fed into the flow, adding to the total volume of water carried. Finally, waste gates had to be constructed to relieve potential overflow and prevent flooding.[94]

Miners had much to consider when constructing a ditch. First, they had to survey the waterway's potential course, which might require a grade of only 0.25 percent, or 12.2 feet per mile. Experts recommended that surveyors employ an aneroid barometer and that the course be leveled and staked. The best ditches were excavated with steep rather than wide, flat sides. The canal's main channel was dug in solid soil; the earth that had been removed was piled on the downside bank. In the Palouse, this step was particularly important because the soil's volcanic ash content reduced cohesion in the downside bank. The problem is illustrated by the story of tough Adam Carrico who saved the ditch leading to his Gold Hill mine by throwing himself prone into a breach after the saturated volcanic soil had given way.[95]

Ditches were dug using a team of horses which plowed the course and turned the earth from the upper side to the lower side of the embankment. One group of miners, the Carricos, used one or more teams in their mining operations and in both ditch and dam construction. Trees were removed from the ditch line by undermining and completely uprooting them. Miners finished their ditches by treading on the lower sides until firm. Once completed, water was turned into the ditch, and after the ground had become saturated, the miners could expect a full flow of the current at their mine site. Afterward, they had to conduct regular maintenance to prevent the downside bank from breaking or flooding.[96]

In the Hoodoo District, mining notices and filings for water rights describe ditches as ranging in capacity from 200 to 2,500 miners' inches. An estimation of the value of ditches with 300 to 3000 miners' inches capacity, considered small by California standards, is available from the Boise Basin rushes. Ditches one to seven miles in length cost $10,000 to $30,000 to construct, and the water they transported cost 60 to 80 cents per inch per 24 hours to use. To insure that these claims were maintained, companies or associations were formed. In the case of the system in White Pine Gulch, for example, a legal division of control was established in 1885 between B.H. and Charles Laughlin, C.C. Roberts, and George M. Wilson.[97]

Dams were built for several purposes. Many were head dams, which collected water from a drainage and directed it into the ditch system. Some were built to create a storage reservoir to insure a water supply during the dry season. Distributing reservoirs needed dams located below the ditch system so that water could be

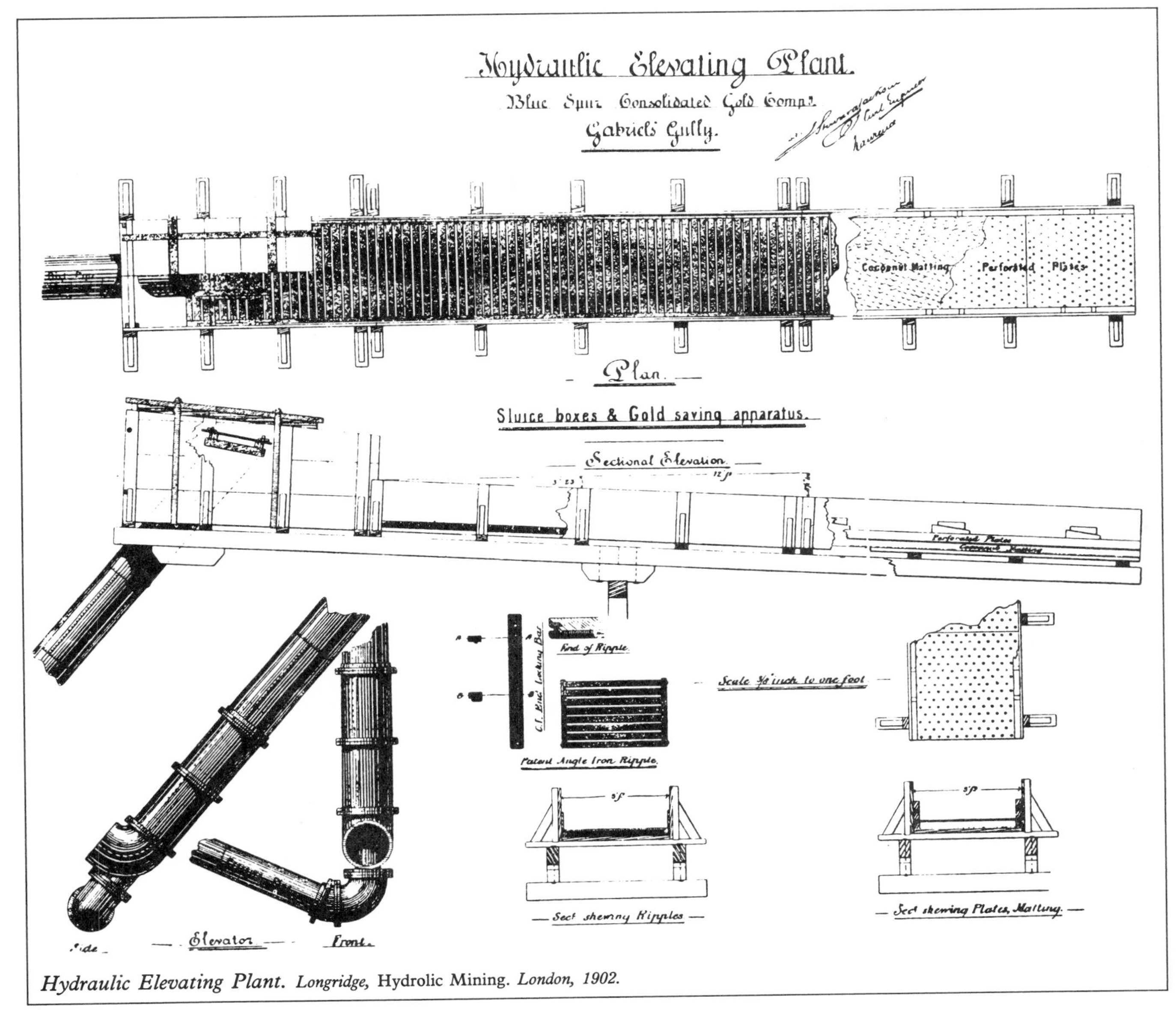

Hydraulic Elevating Plant. *Longridge,* Hydrolic Mining. *London, 1902.*

used on individual claims for a few hours' or days' run. A retaining dam was also better for feeding penstock pipe, rather than risking work stoppages and inconstant supply with feeds directly from ditches. One mining manual stated, "These are especially necessary where the water supply is from mountain streams which have a tendency to slack off in water during the summer months. The erection of a retaining dam for such reservoirs is part of the ditch system." Finally, splash dams were built just above the diggings. These had trip-release gates to suddenly unleash the accumulated water in a powerful flood onto the ground prepared for sluicing.[98]

By and large, the dams in the Hoodoo Mining District were of earthen construction. Publish-

Giant, fitted with deflector, showing elbow base and king-bolt.
Longridge, Hydrolic Mining. *London, 1895.*

ed manuals recommended designs for earthen dams that contained cribbed timber bases. The cribbing was filled with stone, earth, gravel, and sand, and embedded branches of timber were faced upstream to catch future flood sediment, strengthening the dam. The dams were also wider at the bottom, since the force from reservoirs was directed downward to prevent the dams from slipping off their bases. Teams of horses and slipscrapers were employed in dam construction.[99]

Splash dams required special gate construction which would overbalance once the reservoir reached a certain depth. Depending on the size of the dam, the gate might be secured within the earthen wall or be part of a gabion-style wood-and-earth structure. An innovation on the latter form was used by Frank Milbert on the East Fork of Gold Creek. It consisted of a portable steel frame with spring-loaded gate, which insured a sudden release of water. Gates were constructed with the balance axis across the lower one-third. Some miners, however, preferred a trip mechanism to allow them to control the timing of the release.[100]

In general, the best dam sites were locations where the most water could be collected. Reservoirs were filled from the run-off of rain and snow. Steep, barren slopes were recommended to prevent water from escaping. Therefore, hillside vegetation was burned to reduce the evaporation and absorption of moisture.

Dams and ditches were hydraulic structures used for all types of open cut placer mining, of which the most important in the Hoodoo District were ground sluicing and hydraulicking. Ground sluicing was a more controlled version of the ancient practice known as booming. A summary of this method was described by the Roman scholar Pliny, in *Natural History:*

> When they have reached the head of the fall, at the very brow of the mountain, reservoirs are hollowed out a couple of hundred feet in length and breadth, some ten feet in depth. In these reservoirs there are generally five sluices left, about three feet square, so that the moment the reservoir is filled the flood-gates are struck away, and the torrent bursts forth with such a degree of violence as to roll outward any fragments of rock which may obstruct its passage. When they have reached the level ground, too, there is still another labor that awaits

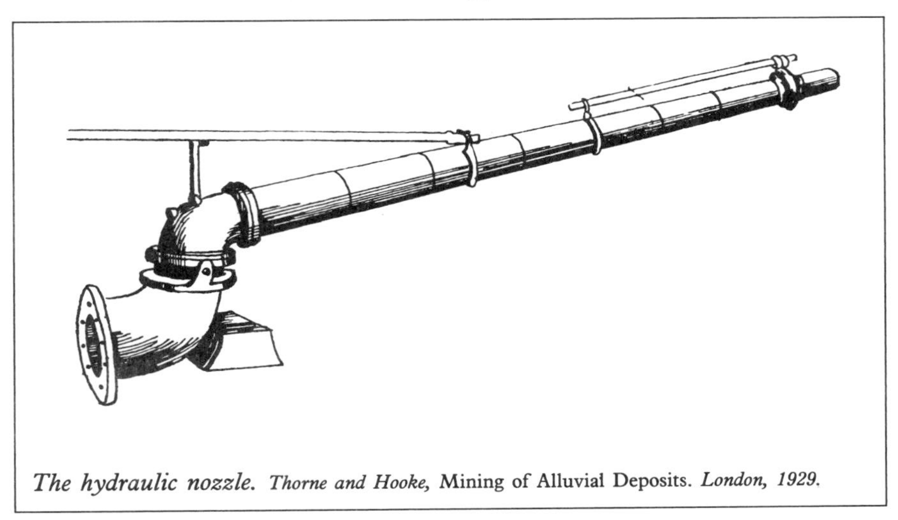

The hydraulic nozzle. Thorne and Hooke, Mining of Alluvial Deposits. *London, 1929.*

them; trenches, known as "agogae," have to be dug for the passage of the water, and these, at regular intervals, have a layer of silex placed at the bottom. This silex is a plant like the rosemary in appearance, rough and prickly, and well adapted for arresting any pieces of gold that may be carried along. The sides, too, are closed in with planks, and are supported by arches when carried over steep and precipitous spots. The earth, carried onwards by the stream, arrives at the sea at last, and thus is the shattered mountain washed away—causes which have greatly tended to extend the shores of Spain by these encroachments on the deep.[101]

Ground sluicing was most useful on steep slopes. It could also be employed on bottomland, but this required careful design of foot dams and tail, or drain races to prevent the downpour of water from reflecting off the tailings and back into the sluice. The sluice itself was cut into the bedrock, using heavy boulders as riffles, and advanced toward the worked bank as it receded. Periodically, the sluice was cleaned up and the accumulated fine material was washed in a sluice box or rocker to obtain the gold. On level land, the ground was sometimes plowed before turning on the water, and a horse-scraper was used to move gravel into the sluice. On slopes, the over-burden was washed away after having been loosened by spading. A gully was then dug, and miners stood on the banks shoveling earth into the flow of water, gradually widening the worked area. In either case, the ground was first cleared of timber and brush.[102]

Hydraulicking was the method of washing down a gravel bank with a high-pressure stream of water directed at the base. It was first used in California to work Tertiary gravels in the central Sierra, but the subsequent industry inundated so much agricultural land with debris that the method became extremely controversial. In Idaho, the use of hydraulic giants and an arrastre on Gold Hill threatened the Cochrane farm on Garden Gulch, and Cochrane filed a damage suit to halt the miners. In the Hoodoo District, the Steffens brothers obtained the giant once owned by the Carricos and used it to work their claim on the North Fork of the Palouse River above White Pine Gulch during the mid-1920s.[103]

"Hydraulic giant" was a generic term which came to be used to describe all nozzles employed to direct water pressure against a gravel bank. The water gained pressure during its gravity fall in large sheet iron pipes, or conduits. The object was to undercut gravels to create controlled

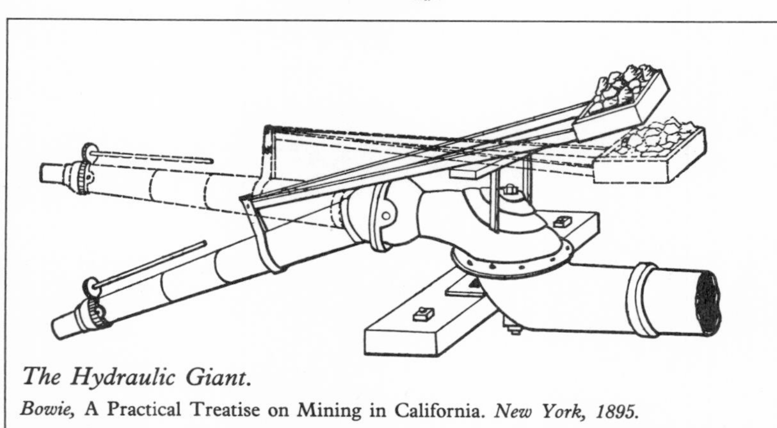

The Hydraulic Giant.
Bowie, A Practical Treatise on Mining in California. *New York, 1895.*

cave-ins and wash the finer material into trenches leading to sluices and undercurrents. Many times more efficient than any hand method, hydraulicking operations could profitably work gravels which paid only three to four cents per cubic yard.[104]

Hydraulic pipe underwent some evolution, especially in California, before the versatility of riveted sheet iron or steel was universally recognized. The pipe was cut in lengths of 30 to 36 inches with diameters of four to six inches which could be transported by mules. The lengths were easily assembled in stovepipe fashion, called "slip joint." It could be singly or doubly cold-riveted along its seam, and the interior was often treated with asphaltum or coal tar to retard corrosion. Jointed together in 20 to 30 foot lengths and sealed with dirt or sawdust, an ordinary single-riveted pipe could withstand 200 pounds of pressure and last 25 years. A slip joint was anchored on steep slopes using hook-shaped lugs on the pipe exterior secured with wire cables. Angle or pressure joints that received greater stress were sealed with lead. Sheet metal sleeves were placed over these joints, and hot lead was poured into the connections.[105]

Hydraulic pipe had several uses at a placer mine. It could serve as a distributing or discharge pipe, or in the pressure box at the head of the system. It was normally not used instead of a ditch, except where a deep gully needed to be crossed or to supply a discharge point down a steep incline. Best results were obtained if the pipe was laid in the shortest, straightest course, because friction caused a loss in "head." The head for a hydraulic system was an indication of the potential pressure, calculated from the perpendicular height of the water fall from entrance to discharge plus the water height above the entrance. It amounted to about one-half pound of pressure per foot of head.[106]

Placer gravels could also be mined by tunneling. This method was called drifting or exploiting, and though there are several adits in the Hoodoo District, there is no evidence that any of these are the result of this type of mining. Drifting was done to reach ancient deep stream gravels so that the mined material could be washed or carted out, or cleaned in sluices built directly within the tunnels.[107]

The most recent method for excavating placer gravels utilized a steam bucket dragline. This heavy equipment simply loaded gravel onto an ore pile or into a gold washing plant. Frank

Milbert at one time contracted to perform this service in Latah County.[108]

The final step in placer mining was the clean-up. Periodically, the miner cleaned the system of washing he was using to obtain the fine material, which held the particles of gold. This was done as frequently as possible to save the maximum amount of gold and yet keep work stoppages to a minimum. Clean-up might be done with a pan, rocker, or any other device, but it was a careful, more detailed job than sluicing the diggings. If mercury was used to amalgamate float or "rusty" gold, the retort process separated the minerals so that the mercury could be reused in the rocker or sluice. If ground sluices caught the concentrates, the use of spoons to clean the surface of the bedrock was called "crevicing." In general, clean-up was especially important because miners were aware of the quantities of gold their sluicing methods failed to recover. Quite often more color was obtained from re-worked tailings than from the original claim. In fact, losses were so great and widespread that the eminent mining engineer and western observer J. Ross Brown noted, "The question arises whether it is not the duty of government to prevent, as far as may be consistent with individual rights, this waste of a common heritage, in which not only ourselves but our posterity are interested."[109]

Laws and Social Life

The gold rushes of the mountain west created a unique social situation of densely populated, compact communities developing rapidly in isolated, wild environments. Despite the boosterism of camp newspapermen, there was little disposition toward permanency. Yet the exploitation of mineral wealth created situations so intense that sometimes only ten feet between claims separated the starving rich from the starving poor. Such conditions required the mediating agencies of a complex institutional organization. The rush to California in 1849 began the evolution of the organization called the mining district, which became an arena for both formal and informal social relationships. This tradition in mining law had its diverse origins in Roman and English civil law; Spanish, German, and Cornish mining experiences; and the committee meetings of the Iowa lead mines. Even as late as 1884 during the Coeur d'Alene rush, the adoption of a district code was a basic task despite the relative proximity of Idaho territorial government and the establishment of a federal mining law:

> While Congress was legislating for one set of conditions and officials journeying slowly over the plains, conventions and elections being held, legislatures gathering and county organizations being effected, the mining population might jump hundreds of miles in a week, fill a gulch with unorganized society, and create conditions imperatively demanding instant readjustment in the application of the focus of government.[110]

In Latah County there were at least nine separate mining districts formed during various periods. As in other areas of the West, districts formed and reformed based upon the community's satisfaction with how internal problems

were solved. Men following the current excitements easily broke away from one district to join another.

It was standard practice for the mining laws adopted in the district miners' meetings to be published in newspapers and county records. The only documentation discovered for the Hoodoo Mining District, however, is the registration of its boundaries by its long-time recorder, W.G. Connor. It reads as follows:

> Hoodoo May 13, 1895
> Jay Woodworth Esq.
>
> Dear Sir:
>
> Yours of the 4" inst at hand to-day, would say in reply the Hoodoo Mining District embraces all of the waters of Palouse River East of Strychnine Creek. inclusion.
> W.G. Connor
>
> Filed for record May 15th 1896 at 4:30 o'clock P.M. request of Jay Woodworth, Rec. fees D.H.
>
> Jay Woodworth
> Recorder, Latah Cty, Idaho

Though this type of evidence is scarce, it reveals the authoritative role the Hoodoo District held in the official view of the duly constituted county government.[111]

At the time of the earliest discoveries in the Hoodoo District, the image of a lone prospector and his mule was largely a myth. Logistical preparation alone required groups ranging in size from five or six to as many as 50 men. Along with mining tools their outfit contained a camp kettle, coffee pot, frying pan, tin cups, and knives; provisions of bacon, beans, self-rising flour, sugar, and coffee; and an armory of revolvers, rifles, and shotguns. Larger parties which set out with some impression of the mineral potential at their destination often organized in advance the methods of claim allocation, settlement of disputes, and district administration.[112]

Until 1872 there was no uniform code for staking placer claims on any of the unencumbered federal lands of the various territories. Territorial legislatures recognized the precedence of regulations established by the mining districts themselves. Local regulations varied regarding the size of claims, but they universally recognized the so-called "prudent man rule." That is, the right of any miner to a particular claim was based upon a minimum amount of development. The reasoning was that if valuable minerals were there, a prudent man would not fail to work for them. As a result of pressure by mining interests who wished to secure mineral titles for capital development, and the concern in government over federal control of public lands, Congress passed the 1866 mining law. The law stated that federal lands were open for exploration and occupation, and it described the circumstances under which mineral rights could be obtained. Importantly, it also recognized that the miners in mining districts regulated themselves. This law provided the legal basis of resolution for innumerable individual litigations. The ensuing judicial commotion pointed out weaknesses in the law which were corrected in the 1872 revision. The 1872 law specified the amount of work necessary to hold a claim, the information which was to appear on every location notice, and required that claim boundaries be clearly marked.[113]

The location notices themselves had a common form with a description of the claim and its boundaries followed by the testament of a witness

that the claim had indeed been duly located. Claim boundaries were marked in numerous ways, but the most frequent methods were to blaze trees or square a tall stump, oriented to the points of a compass, four to 12 inches in diameter and about five feet high. Boundary information was then written on the blazes or stump faces, and a copy of the location notice was appended to the discovery shaft monument, usually placed inside a tin tobacco can nailed to a tree. In 1902, the filed location notices began to make reference to township and range map coordinates and to rely less upon common landmark names.[114]

After claims were staked and the business of mining begun, the establishment of a mining camp or town quickly occurred. Two factors which helped define camps—besides the notorious entertainments provided—were the need for a place to accomplish district business and hold meetings, and the dependence of miners upon supply entrepots. In addition to Hoodoo and Grizzle Camps, there were other locations which received passing notice as congregating places in the Hoodoo District: Long Jim Lockridge's cabin at the mouth of Jerome Creek; the Havner store-boarding house on the North Fork of the Palouse River; various cabins of the mining district recorders; Jake Johnson's halfway house at Woodfell; and a place referred to in two location notices by Oliver Hazard Perry Beagle with the unsavory title of Dogtown, "1½ miles S from Strickims Butts." In 1888, John Banks leased his claim in Blaine Gulch to a Chinese, Sam Jon, along with the use of ditches, water rights, sluice boxes, a cabin, and a store. In 1913, the location was still known as the old store, though Blaine Gulch had become Banks Gulch.[115]

Little is known about the early miners' meetings held in the northern Latah County districts, but the Idaho Territory recognized their quasi-governmental functions. County recorders were required to appoint a deputy mining recorder at any necessary place or a place with at least ten locators. If the county recorder failed to do so, the miners could elect one of their number to hold the office until such time as the appointment was made. Miners' meetings were still being held on Gold Hill in the 1930s and 1940s, but little in the way of formal business was transacted. Claimjumpers, sluice-robbers, and "sharp practices" continued to receive judicial hearing, and claims were defended with firearms in theory if not in fact. The small, close-knit community gathered more frequently, however, to swap news and stories and listen to "Death Valley Days" on Park Shattuck's radio.[116]

The district mining recorder served at the behest of his peers and without salary. He did receive a portion of the claim notice filing fee, about three dollars, out of which he had to travel to the county courthouse to register the claims recorded. Abuses of the office most frequently occurred in collecting fees and taxes from Chinese. Unprotected by public justice in the miners' meeting until lower wages drove most whites away, the Chinese were not a part of the miners' social or legal systems. An Idaho tax was passed charging foreigners five dollars per month, but its wording was such that it applied only to Chinese. Chinese miners were forced to pay the exceptional tax and were sometimes robbed by officials under the guise of "watchmen" and "collectors."[117]

As to the second factor of miners' dependence upon supply entrepots, packers and settlers who handled miners' goods constituted the earliest Euro-American trade network in the Palouse.

Generally, a freighter used a large wagon with smaller trailers pulled by mules, often with much makeshift equipment to facilitate travel over unimproved trails. Some sold goods off the tailgate, but most sources state that Hoodoo miners sent specific orders to Lewiston or Walla Walla to be filled and delivered to some point in the upper Palouse Valley. Homesteader Matthew Miller at Kennedy Ford acted as an agent for miners, storing supplies in his log warehouse until called for by the miners. Miller's services were especially important to Chinese miners because he was an intermediary supply agent able to avoid discriminatory pricing. During later periods, Hawkins's and Starner's stores at present-day Hampton catered to miners' needs.[118]

Storekeepers and packers were often partners in trade to the mines. The precise nature of these arrangements for shipping to the Hoodoo District is unknown, but several kinds of ties seem likely. Both C.H. Farnsworth's and J.C. Northrup's livery stables ran freight lines and stages to the Hoodoo mines. J.G. Powers sent gold dust to Spokane from the Bank of Palouse City, where J.A. Starner, Daniel Preffer, Jesse Bishop, Ed Cheney, A.A. Kincaid, and Farnsworth also held depositor accounts. Ankcorn Hardware in Palouse carried mining tools, and F.H. Ankcorn, the proprietor, had interests with minerals speculators W.R. Belvail and W.F. Chalenor. Correspondence between W.C. Wells, of ex-miner E.K. Parker's General Merchandise in Princeton, and J.P. Duke, Palouse businessman, called for greater cooperation between the towns in economic ventures.[119]

Miners often bemoaned the inferior quality of goods they received and the exorbitant prices they paid. They blamed eastern merchants for preying on the isolated camps, which had no choice but to accept what was made available. Bulk shipments continued to plague local merchants as well, especially when mail order houses began to provide competition. Suppliers would ship an unsolicited consignment to their retailers, and the retailer had to either sell the lot or pay the return freight charges. By contrast, Montgomery Ward or Sears and Roebuck offered a line of miners' spades, sluice forks, and drain spades of uniform quality, and deliverable to the nearest post office.[120]

Early residents of the upper Palouse Valley periodically made the 30-mile trip to Palouse City by wagon, buggy, bicycle, horseback, or on foot. Besides obtaining necessary clothing and provisions, there were also appointments with the dentist and days at the circus. Travel also occurred in the opposite direction, as settlers made excursions up the Palouse River to the Hoodoos to picnic, pick huckleberries, and observe the activities during any of the gold excitements. Some miners paid the boys of homestead families to bring groceries, cut wood, and do the annual assessment work on their claims.[121]

However, Hoodoo miners relied more on packers who could bring supplies over the Hoodoo Trail. The best-known supply route was used by the best-known packer, Jake Johnson. It trailed from his halfway house and farm at Woodfell up Pup Creek, across the ridge to Excavation Gulch, down to Strychnine Creek, across Strychnine Ridge, and down into the Poorman Creek canyon. Typically, Johnson farmed less than one-third of his timber homestead, and for additional income packed miners' provisions with his three horses. Every two weeks he took a two-day round-trip by wagon to Palouse, reloaded the groceries onto the horses at Woodfell, and then made the one-day trip to the mines. He

continued to supply the miners after 1905, using the newly constructed wagon road.[122]

The lives of Hoodoo miners were lonely, but many liked it that way. Some, living within a quarter mile of one another, visited each other only once a year. Some stayed many years, going to the mines as young men and dying there. With independence as a central characteristic of their personalities, most were bachelors. Of the resident population of about 20 along the North Fork of the Palouse River after 1910, only three were married: John English, Jack Connor, and Gene Wiesenthal.[123]

While some miners constructed log cabins, most sought shelter in semi-subterranean lean-tos, about six feet by eight feet in size. These were made from split logs and had a fireplace of mud and rocks and a bed of piled brush. Along with bacon and beans, meals consisted of produce obtained from homesteaders' and Chinese truck gardens. Condensed milk was the most important canned food. Eggs could be preserved in paraffin or mineral oil if they were frequently turned so that the yolk did not touch the shell. Lemons were kept in a glass jar and wiped off when "sweaty." There was no need for a mixing bowl, since flapjack batter could be mixed in the top of the flour sack. These simple needs meant that a miners' cabin need only be shelter from the elements, and they reduced to a minimum his critical dependence on expensive outside suppliers and exorbitant freight lines.[124]

Conclusion
Grubstaking the Palouse

In all, perhaps as much as $1,250,000 worth of gold (measured in 1980s currency) was taken from the Hoodoo Mining District. That is insignificant compared to the histories of the major mining rushes of the West. The economic impact must be further discounted by realizing that roughly half of that figure represents the recovery netted by dredging the North Fork Palouse River during the 1940s. Why should we be concerned over such a forgotten episode in north Idaho regional history?

The answer lies in understanding the historic changes in the Palouse River Valley communities. Without the numerous, nameless packers who toiled out of the Snake River canyon lugging supplies, the legendary aura around a gold discovery by someone named Hoteling would not exist. Without the hundreds of miners laboring over dozens of claims, the enthusiastic boosterism of Palouse City businessmen would have had a hollow ring. Without a steady, nearby consumer market, the faith of the early farmers in the agricultural potential of their lands would have been misplaced.

Throughout this history of the Hoodoo Mining District, individuals gained personal success because of their relationships with a community. In fact, our historic definition of the pioneer spirit recalls individuals overcoming extraordinary obstacles in order to realize a vision of security and society. Yet these myths and symbols about the past must by tempered with the understanding that individuals and communities grow together. Individuality without community is a poor condition.

When a miner went into the hills to develop his claim, he took along what he called his "grubstake." This was the investment, his own or someone else's, in equipment and supplies that was to carry him through until some dreams of profit could be realized. The grubstake was a good risk partially because the miner knew that the community and customs of the mining district would protect it and insure that it would be completely devoted to his own claim. His individual rights were guaranteed as long as he was part of the community. His individual success as a miner, however lonely the work in the hills might become, was due to the social relations of the district community.

The symbol of the grubstake applies to the relationship between the Hoodoo Mining District and the communities of the Palouse River Valley. Periodic exploitation of placer gold injected some real wealth into the growing regional economy and helped develop commitments to building farms and businesses. Mining the gold required a cooperative effort by the members of the communities, and that effort can not be measured by the dollars-and-cents value of bullion. The communities of the Palouse have more than redeemed their grubstake.

Glossary

Many general works on placer mining history and technology include comprehensive glossaries, so the following list is intended only as an immediate aid for terms used in this work. It is drawn primarily from Mary Leucke, Brian Power, and Jim Rock, "A Glossary of Mining Terms," manuscript report U.S. Department of Agriculture, Region 5, Klamath National Forest, and Otis E. Young, Jr., *Western Mining* (Norman: University of Oklahoma Press, 1970), pp. 293-317. Other terms defined had specific use in the Hoodoo District.

Adit: A mostly horizontal passage driven from the surface for the working of a mine. An adit has only one opening, as distinguished from a tunnel.

Amalgam: An alloy of mercury with gold or another metal. In the case of placer gold, a "dry" amalgam, one from which all excess mercury has been removed by squeezing through chamois leather will contain nearly equal proportions of gold and mercury.

Arrastre: A circular mill for grinding quartz by trituration between stones attached loosely to cross arms.

Assay: To determine the amount of metal contained in an ore.

Assessment work: Annual work done upon an unpatented mining claim necessary for the maintenance of the possessory title.

Bench placer: Gravel deposits in ancient stream channels and flood plains which stand from 50 to several hundred feet above the present streams.

Cribbing: Close timbering, as the lining of a shaft. In placer work, cribbing may be needed to support the walls of shaft or test pit put down in loose or wet ground.

Drag: The finer, heavier sands left in the pan after each swirl of water and gravel.

Drift: A sub-tunnel run from the main tunnel to prospect for the pay lead or block out the ground to facilitate its working.

Float: Small and thin particles of gold which have not been transported far from the vein source.

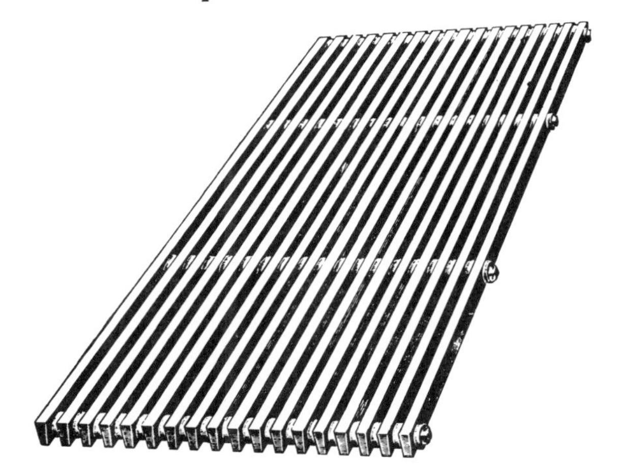

The grizzley. Longridge, Hydrolic Mining, *London, 1895.*

Grizzly: A grating, usually iron, which serves as a heavy-duty screen to prevent large rocks or boulders from entering a sluice or other recovery equipment.

Ground sluicing: A mining method in which the gravel is excavated by water not under pressure. A natural or artificial water channel is used to start the operation and while a stream of water is directed through the channel or cut, the adjacent gravel banks are brought down by picking at the base of the bank and by directing the water flow as to undercut the bank and aid in its caving. Sluice boxes may or may not be used. Where not used, the gold is allowed to accumulate on the bedrock awaiting subsequent clean-up. A substantial water flow and bedrock grade are necessary.

Hillside placer: A group of gravel deposits intermediate between the creek and bench placers. Their bedrock is slightly above the creek bed, and the surface topography shows no indication of benching.

Hornspoon: A shallow, oblong vessel, at one time made from a section of ox horn but now made of metal. Used to test small samples of gold-bearing material by washing, in a manner similar to panning.

Hydraulic mining: A method of mining in which a bank of gold-bearing earth or gravel is washed away by a powerful jet of water and carried into sluices, where the gold separates from the earth by its specific gravity.

Location: A validly registered mining claim which has been shown to contain a valuable mineral deposit.

Ore: Metallic minerals in concentrations that can be worked at a profit.

Penstock Overflow, showing sand catch, screen, overflow and emptying sluice.

Penstock: A conduit or pipe for conducting water; or a gate for regulating water flow.

Race: In placer mining, a narrow watercourse used to direct a strong flow of water; especially as the drain or tail races at the foot of ground sluices.

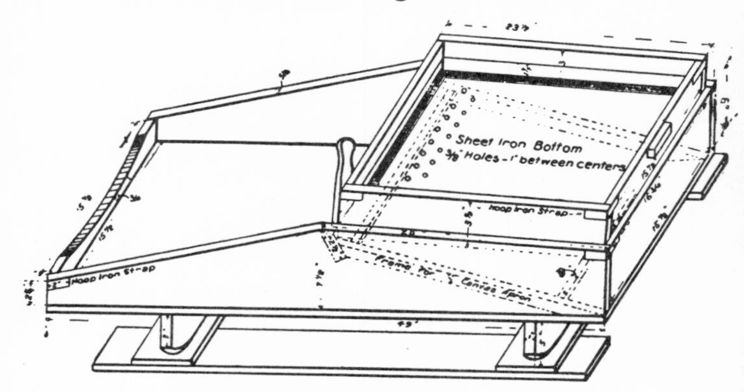

A typical rocker. *Longridge,* Hydrolic Mining. *London, 1895.*

Rocker: A short, sluice-like trough fitted with transverse curved supports, permitting it to be rocked from side to side, and provided with a shallow hopper at its upper end.

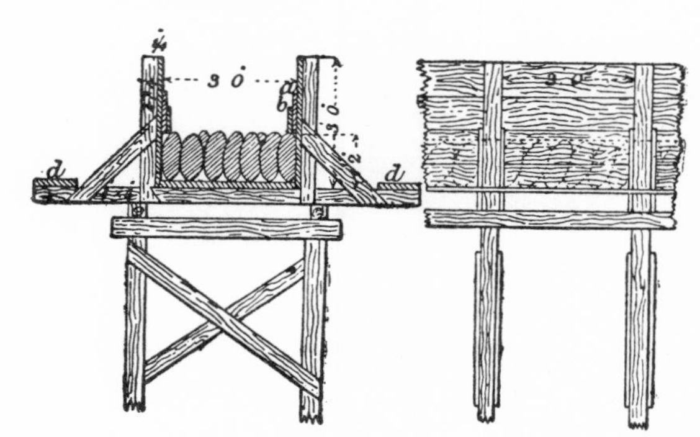

Rock-paved sluice on trestles. *Longridge,* Hydrolic Mining. *London, 1902.*

Sluice box: An elongated wooden or metal trough, equipped with riffles, through which alluvial material is washed to recover its gold or other heavy minerals.

Stamps: Machinery for crushing ore with the presence of water and heavy iron blocks.

Stope: An excavation from which ore has been taken in a series of steps. Usually applied to highly inclined or vertical veins.

Wood block riffles. Longridge, Hydrolic Mining, *London, 1895.*

Washings: The ore undergoes occasionally two or three washings; the first process being that of washing the slime and earthy particles from the rougher and larger stones of ore.

Waste: Valueless material such as barren gravel or overburden. Sometimes used interchangeably with "tailings," as material regarded as too poor to be treated further.

Notes

1. For the district's boundaries see John N. Faick, "Geology and Ore Deposits of the Gold Hill District," unpublished master's thesis, University of Idaho, 1937, p. 52; *An Illustrated History of North Idaho* (San Francisco: Western Historical Publishers, 1903), p. 585; Albert Russell McNeill, III, "Geology of the Hoodoo Mining District, Latah County, Idaho," unpublished master's thesis, University of Idaho, 1971, p. 54; John B. Miller, *The Trees Grew Tall* (Moscow, Idaho: News-Review Publishing Co., 1972), p. 19; and Harold R. Boyd, "Terror in the East Palouse," *The Pacific Northwesterner*, 2:3 (1958), p. 34.

2. The north-south route earlier favored by area Indians closely approximates that followed by modern Highway 95. It was a route also used by early missionaries, especially Fathers Pierre DeSmet and Joseph Cataldo. See Kenneth B. Platt, *Some Pioneer Glimpses of Latah County* (Moscow, Idaho: Latah County Historical Society, 1974), p. 6; Johnson Parker, "Environment and Forest Distribution of the Palouse Range in Northern Idaho," *Ecology*, 33:4 (1952), p. 38; Frank Milbert, "Men of Gold Creek," unpublished manuscript in the resources culture history file, Palouse Ranger District, Potlatch, Idaho, p. 8; and the Palouse Story (Palouse, Washington: Palouse Town and Country Study Program, Report of the History Committee, 1962), p. 10.

3. Sylvia Ross and Carl N. Savage, *Idaho Earth Science*, Earth Science Series No. 1 (Boise: Idaho Bureau of Mines and Geology, 1967), pp. 136, 143-44; Don J. Easterbrook and David A. Rahm, *Landforms of Washington, the Geologic Environment* (Bellingham, Washington: Union Printing, 1970), pp. 118, 153: *General Soil Survey, Latah County, Idaho* (U.S. Soil Conservation Service, 1973), pp. 3-4, 8-9.

4. Measurement of the seasonal discharge by the Palouse River and several of its tributaries in the Hoodoo Mining District began annually in 1974. The results indicate that peak flow can be expected in mid-to late-April, and that the flood stage is dramatically higher than low water, as might be expected from snow-fed mountain streams. The low flow occurs in early-to mid-September and again in mid-to late-February. Thus, not only are the streams low after the summer season, but the cold temperatures and snowpack hold the increased moisture of late winter. This relationship accentuates the dramatic effect of the late April flood stage. See Water Uses and Development, Palouse River 1974-77, File 2540, Palouse Ranger District, Potlatch, Idaho.

5. E.N. Franq, "An Ecologic Study of Mammals of the Palouse Range," unpublished master's thesis, University of Idaho, 1962, p. 9; Richard Waldbauer, *Poorman Creek Cost/Share Road*, Cultural Resource Management Project #80-PAL-4 (Potlatch, Idaho: Clearwater National Forest, Palouse Ranger District, 1980).

6. A.L. Anderson, "A Copper Deposit of the Ducktown Type near the Coeur d'Alene District, Idaho," *Economic Geology*, (1941), pp. 641-57; Faick, "Geology of the Gold Hill District;" McNeill, "Geology of the Hoodoo District."

7. Though the source for gold in the Hoodoo District has not been identified, both A.L. Anderson and Albert McNeill postulate igneous activity in the Belt Supergroup rocks as the probable cause sequence. McNeill states that "Metalization occurred during one or more of the metamorphic events, which appears to have caused remobilization of metals from the Belt Supergroup rocks into zones of accumulation such as the Mizpah fault."
   The belt rocks themselves are metasedimentary and metamorphic. The younger Striped Peak Formation consists of two members composed of thin-bed shales and quartzites. These occur around Bald Mountain and Little Bald and above White Pine Gulch on the North Fork Palouse River. Toward the southeast and along the lower reaches of the North Fork are the three members of the older Wallace Formation, which is composed of thin-bedded, fine-grained calcareous quartzites, impure limestones, argillites, and shale. The crucial events for metals formation were the periods of folding and faulting of Belt rocks. The two major features are the Hoodoo Fault, which tends northwest-southeast along the crest of the Hoodoo Mountains, and the Mizpah Fault, which tends southwest-northeast and upon which the Mizpah copper mine is centered. It is the incompletely understood radial fractures centered at the head of White Pine Gulch, however, which are associated with gold quartz veins. This conclusion is corroborated by similar patterning at Gold Hill, twelve miles to the west. At Gold

Hill, mineral-bearing solutions from the Belt rocks were remobilized along a temperature-pressure gradient; the resulting metals were contained by zones of accumulation, such as faults.

In closely examining this radial fault pattern, the entire bed of White Pine Gulch occurs upon one of the supposed concealed arms. Closely associated are the bed of Moscow Gulch and a precisely located fault that runs southeast, across the divide between White Pine Gulch, through the North Fork Palouse River, and through the northeast corner of Sec. 12 T42N R2W. This second location is especially notable as the site of one of the prominent hillside placers called "China Hill."

There are also several faults, unrelated to the anomalous radial pattern, which are important as locations for historic placer activity. Mountain Gulch, California Gulch, Banks Gulch, the forks of the Palouse River between Beagle Gulch and just above White Pine Gulch are all mapped as concealed fault locations. Hillside and gulch placers occurred on faults on sections 22, 15, 10 and 11, T4N, R2W.

Finally, McNeill notes that there is an "abrupt change in the lithologic character of the rocks between the two formations (Wallace and Striped Peak) [which] suggests an unconformity of some magnitude." The Striped Peak Formation represents a general lack of metamorphism, while sequences of metamorphism have created the schists and quartzites at the top of the Wallace formation. Since there is a lack of gradation from high-grade contact metamorphism in the Wallace formation to the Striped Peak Formation, it is thought that a considerable period of deposition is missing from the stratigraphic record. No estimation has been made about any influences this deposition may have had on Palouse River alluvium or ancient streambeds on high mountain slopes.

See Anderson, "A Copper Deposit of the Ducktown Type," pp. 655-56; McNeill, "Geology of the Hoodoo District," pp. 11, 19, 21, 52, 57-8, 95 and Plate I; and Faick, "Geology of the Gold Hill District," p. 41. For the China Hill mine see Latah County Mining Notices, Book 1, p. 454; Book 2, pp. 13, 203 and 207.

8. Ralph Burcham, J., "Reminiscences of E. D. Pierce, Discoverer of Gold in Idaho," Ph.D. dissertation, State College of Washington, 1958, p. 15.

9. Fort "Colville" is the name of the United States Army fort; Fort "Colvile" is the name of the Hudson's Bay Company post. For a discussion of the names and how they are easily confused, see Craig E. Holstine and Fred C. Bohm, *The People's History of Stevens County, Washington* (Colville, Washington: Stevens County Historical Society, 1983), pp.6-8, 13-15.

10. Subsequently, the ebb and flow of populations between the Pacific Northwest and California established a complex economic pattern that alternately caused the economies of the two regions to become subsidiary to one another in providing mining supplies and equipment. See Dorothy O. Johansen and Charles M. Gates, *Empire of the Columbia: A History of the Pacific Northwest* (New York: Harper & Row, 1967), p. 265; William J. Trimble, *Mining Advance into the Inland Empire*, Bulletin No. 638 (Madison: University of Wisconsin, 1914) p. 16; Elaine Tanner, "A Study of the Underlying Causes of the Depression of 1854," Reed College Bulletin, 25:3 (1947), p. 60; and Burcham, "E. D. Pierce," p. 30.

11. The best sources for the Indian Wars of 1855-58 are U.S. government documents. See especially 34th Cong., 1st Sess., H. Ex. Docs. 93 and 118 (1856); 34th Cong. 3rd. Sess., H. Ex. Doc. 1 (1856), pp. 147-203; and J. Ross Browne, *Indian Wars in Oregon and Washington Territories*, 35th Cong., 1st Sess., H. Ex. Doc. 38 (1857-58). Among the more reliable secondary accounts are D. W. Meinig, *The Great Columbia Plain: A Historical Geography, 1805-1910* (Seattle: University of Washington Press, 1968), pp. 152-68; Kent D. Richards, *Isaac I. Stevens: A Young Man in a Hurry* (Provo: Brigham Young University Press, 1979); Richard Scheuerman and Clifford Trafzer, "The First People of the Palouse Country," *The Bunchgrass Historian*, 8:3 (1980), pp. 3-18; and Richard Scheuerman and Clifford Trafzer, *Renegade Tribe: The Palouse Indians and the Invasion of the Inland Pacific Northwest* (Pullman: Washington State University Press, 1986).

12. See Herman J. Deutsch, "The Evolution of the International Boundary in the Inland Empire of the Pacific Northwest," *Pacific Northwest Quarterly*, 51:2 (1960), pp. 49-56; and "A Contemporary Report on the 49 Boundary Survey," *Pacific Northwest Quarterly*, 53: 1 (1962), pp. 17-33. Also see Trimble, "Mining Advance into Inland Empire," pp. 27-8, 37-8.

13. Trimble, "Mining Advance into Inland Empire," p. 37.

14. Merril D. Beal and Merle W. Wells, *History of Idaho* (New York: Lewis Historical Publishing, 1959), pp. 281-88, 298; Byron Defenbach, *Idaho: The Place and Its People* (New York: American Historical Society, 1933), p. 258; Hiram T. French, *History of Idaho* (Chicago: Lewis Publishing Co., 1914), p. 26; Lawrence R. Harker, "A History of Latah County to 1900," unpublished master's thesis, University of Idaho, 1941, p. 38; James H. Hawley, *History of Idaho, The Gem of the Mountains* (Chicago: S. J. Clarke Co., 1920), pp. 103-4; *Illustrated History of North Idaho,* p. 19; Burcham, "E. D. Pierce," p. 32.

15. John Mullan, *Report on the Construction of a Military Road from Fort Walla Walla to Fort Benton,* 37th Cong., 3rd Sess., S. Ex. Doc. 43, p. 13.

16. Otis E. Young, Jr., *Western Mining* (Norman: University of Oklahoma Press, 1970), p. 16.

17. Mullan, *Report on Military Road,* p. 97. The Coeur d'Alene Indian name for the Palouse Range is Tat-hu-nah.

18. Ibid., pp. 97, 99.

19. Ibid., p. 104; Isaac I. Stevens, *Explorations for a Railroad Route from the Mississippi River to the Pacific,* vol. 12, part I (Washington: Thos. H. Ford, 1860), pp. 199-200.

20. Faick, "Geology of the Gold Hill District," p. 52; C. R. Hubbard, *Mineral Resources of Latah County, Idaho, County Report No. 2* (Moscow: University of Idaho; Idaho Bureau of Mines and Geology, 1957), p. 10; G. D. Kincaid, *Palouse in the Making* (Rosalia, Washington: *The Citizen Journal,* (reprint, 1979), p. 13; Miller, *Trees Grew Tall,* p. 9; Milbert, "Men of Gold Creek," p. 105; *History of North Idaho,* p. 585; Glen Palmer, "The Hoodoo Trail" oral history interview transcript in the resources culture history file, Palouse Ranger District, Potlatch, Idaho, p. 1; Latah County Mining Notices, Book 4, pp. 216-19; Nez Perce County Final Receipts, Book C, Section D, pp. 60-1.

21. Paul T. Bockmier, Jr. Papers, Washington State University Manuscripts, Archives, and Special Collections, Cage 94, Folder 10; Will E. Wiley, "Gold Districts of the Northwest, with a Special Study of the Hoodoo District of Northern Idaho," 1915, typescript in the Pacific Northwest Collection, University of Washington Library, p. 10; Milo Wesley Goss, transcripts of interview with G. D. Kincaid for "The Northwest Pioneer," series of KWSU radio broadcasts in the 1930s, Washington State University, Manuscripts, Archives, and Special Collections, Cage 1566.

22. *History of North Idaho,* pp. 585-6; Boyd, "Terror in the East Palouse," p. 34; Alton B. Oviatt, "Pacific Coast Competition for the Gold Camp Trade of Montana," *Pacific Northwest Quarterly,* 56:4 (1965), p. 168; Alec L. Bull, "Palouse Valley History," typescript in the resources culture history file, Palouse Ranger District, Potlatch, Idaho, p. 3. Bull's manuscript was published in *Latah Legacy, the Quarterly Journal of the Latah County Historical Society,* 9:1 (1980), pp. 9-16.

23. Bull, "Palouse Valley History," p. 4; Milbert, "Men of Gold Creek," p. 2; Bockmier Papers, Folder 10; Faick, "Geology of the Gold Hill District," p. 52; Hubbard, *Mineral Resources of Latah County,* p. 10; *History of North Idaho,* p. 586.

24. General Land Office Map, December 1874, T41N R4W BM.

25. *The Palouse Story,* pp. 11-12; *History of North Idaho,* p. 581; Kincaid, *Palouse in the Making,* p. 1; Roy Buhl McKinney, interviewed by Margot Knight, 13 December 1977, Whitman County Historical Society oral history collection. Palouse soil was not an inducement to early immigration because of the abundance of what was thought to be better land throughout the rest of the West. The altitude and low annual rainfall on the plateau were also thought to make the region unprofitable for agricultural enterprise. "Very few came to locate with a view of establishing a home here, their purpose being to graze stock for a few years and then abandon the country, raising some grain in the meantime for their own use, and possibly a little to sell, if anyone wished to buy," noted Frank T. Gilbert, *Historic Sketches of Walla*

Walla, Washington, Columbia and Garfield Counties, Washington Territory (Portland: A. G. Walling Co., 1882), p. 224. Also see pp. 442-43. Later, Ewing and Atwood, as ranchers refused to sell cattle to the "nesters"—as homesteaders were called. In Ewing's case, he declared that he had come to the Palouse to get rich, and he did not believe he could do so by selling to immigrants.

26. Gilbert, *Historic Sketches*, p. 433; Newton C. Abbott and Fred K. Carver, *The Evolution of Washington Counties* (Yakima, Washington: Yakima Valley Genealogical Society and Klickitat County Historical Society, 1978), pp. 61, 63.

27. *The Palouse Story*, p. 11; Frank R. Schell, *Ghost Towns and Live Ones: A Chronology of the Post Office Department in Idaho, 1861-1873* (Frank Schell, 1973), p. 24. Also see Lalia P. Boone, "Post Offices of Latah County," *Latah Legacy, The Quarterly Journal of the Latah County Historical Society*, 7:4 (1978), pp. 3-29; and for a general overview of winter cattle kills in the Northwest, J. Orin Oliphant, "Winter Losses of Cattle in the Oregon Country, 1847-1890," *Washington Historical Quarterly*, 32:1 (1932), pp. 3-17.

28. Earlier, in 1872, Truax ran a post office. Gilbert, *Historic Sketches*, pp. 443-4.

29. Goss, interview with Kincaid; Kincaid, *Palouse in the Making*, p. 14; Wiley, "Gold Districts of the Northwest," p. 11.

30. Gilbert, *Historic Sketches*, p. 446; *The Palouse Story*, pp. 11-12, 19; *History of North Idaho*, p. 582.

31. Gilbert, *Historic Sketches*, p. 445.

32. Nez Perce County Commissioners' Record, Book B, Section A, p. 15; Nez Perce County Road Book, No. 2, p. 105; *History of North Idaho*, p. 582.

33. The quotation is from Verna Palmer Hardt, transcript of oral history interview in Latah County Historical Library, p. 4. Also see Bull, "Palouse Valley History," p. 4; Palmer, "The Hoodoo Trail," p. 1; and Keith Petersen and Mary E. Reed, "Latah Vignettes: Laird Park," *Latah Legacy, The Quarterly Journal of the Latah County Historical Society*, 13:1 (1984), pp. 20-2.

34. Wiley, "Gold Districts of the Northwest," p. 10; Hardt oral history transcript, p. 27; General Land Office Map, December 1874, T41N R3W BM; Nez Perce County Final Receipts Book C, No. 2, p. 391.

35. *The Palouse Story*, pp. 21, 23; Gilbert, *Historical Sketches*, p. 443; Kincaid, *Palouse in the Making*, p. 1. For more on the significance of Palouse as a sawmilling center see Robert W. Swanson, "A History of Logging and Lumbering on the Palouse Rivers, 1870-1905," unpublished master's thesis, State College of Washington, 1958.

36. *The Palouse Story*, p. 20, "Quite a large amount of gold was being mined from the Hoodoo Mountains, and there is where a lot of Chinese headed for. Several pioneers conducted a business of transporting people to the mines."

37. The quotation is from the Hardt oral history transcript, pp. 28-9. Also see *The Palouse Story*, p. 20. For a discussion of the origins and evolution of local geographical names see Boone, *From A to Z in Latah County, Idaho: A Place Name Dictionary* (Moscow: Lalia Boone, 1983), p. 42; and Petersen and Reed, "Laird Park."

38. A.A. Bensel, *Bensel's Palouse City Directory for the Year 1891* (Fond Du Lac, Wisconsin: A.A. Bensel, 1891), pp. 29-30; Nez Perce County Deeds, Book N, pp. 167, 184.

39. Milbert, "Men of Gold Creek," p. 9; Miller, *Trees Grew Tall*, p. 23; *Palouse News*, 30 Apr. 1893, 25 Aug. 1893.

40. The quotations are from the *Palouse Republican*, 23 Oct. 1893; and the *Palouse News*, 15 Apr. 1892. See the *Moscow Mirror*, 16 Oct. 1886 for a typical local story concerning Chinese immigration.

41. Kincaid, *Palouse in the Making*, p. 15; Wiley, "Gold Districts of the Northwest," p. 11; C.T. Stranahan, *Pioneer Stories* (Lewiston, Idaho: Lewiston Chapter, Idaho Writer's League, 1947), p. 12; John P. Esvelt, "Upper Columbia Chinese Placering," *The Pacific Northwesterner*, 3:1 (1959); pp. 6-11; *Moscow Mirror*, 11 Dec. 1891.

42. Those filing the claim were Andrew Grube; Ed Graham, a homesteader from Meadow Creek; E.E. James; F.M. Smith, proprietor of the Black Hawk Livery Stable; John Wood; John Malhern, owner of the Cozy Saloon; and G.C. Havner, who is said to have established a town on the North Fork Palouse River which boasted a saloon, livery stable, and blacksmith shop. See Nez Perce County Mining Notices, Book P, p. 251; Palouse News, June 5,1884; General Land Office Map, 1883, T42N R3W BM; Wiley, "Gold Districts of the Northwest," p. 10.

43. Among Palouse City businessmen who established claims were: Cy Roberts, miner; Jacob Slaght, blacksmith; John Banks, real estate man; E.J. Cheney, treasurer of the Palouse Mercantile Company; J.C. Northrup, real estate man; George Henwood, postmaster and dentist; E.H. Orcutt, editor of the Boomerang and a claims collector; Daniel Preffer, City Hotel owner; W.S. Reider, drayman at the Blackhawk Stables; and Frank Truett, druggist. The upper Palouse Valley homesteaders were D.C. Tribble and John G. Hoskins, Sr.

44. Wiley, "Gold Districts of the Northwest," p. 11; Nez Perce County Mining Notices, Book P, pp. 254-5; *Bensel's Palouse City Directory for 1891*; Palouse *Boomerang*, July 4, 1883; *Palouse News*, June 5, 1884.

45. *Palouse News*, June 5, 1884.

46. Nez Perce County Mining Notices, Book P, pp. 254, 255.

47. Nez Perce County Mining Notices, Book N, p. 30.

48. John Wood, Robert Dixon, Charles Reitz, B.H. Laughlin, Charles Laughlin, Frank Points, T.H. Wadsworth, A.A. Smith, F.C. Smith, John Banks, and Chris Stanull all deeded mining ground to the Chinese at the rate of approximately $100 per twenty acres. The claims were in the vicinity of Cleveland and Banks Gulches. The legal documentation was handled by Palouse City brokers, A.A. Kincaid, J.G. Powers, J.W. Breeding, and J.H. Wiley. The remainder of the claims were probably prospected by area residents to provide extra income after harvest. Nez Perce County Deeds, Book N, pp. 164-209.

49. Unfortunately, only one of Conner's volumes—for the years 1889-1894—is known to still exist and is in private possession. A photocopy of this book can be found in the Latah County Historical Society Library, Archives SC/CON-1. Also see Nez Perce County Mining Notices, Book N. pp. 134, 136, 200.

50. The quotation is from *The Palouse Story*, p. 90. Also see Gilbert, *Historic Sketches*, p. 443; *Bensel's Palouse City Directory for 1891*, p. 11; *Palouse News*, 5 May 1893; D.E. Livingston-Little, *An Economic History of North Idaho, 1800-1900* (Los Angeles: Journal of the West, 1965), p. 80; Swansen, "A History of Logging and Lumbering on the Palouse Rivers," and Keith Petersen, "Farewell to the Potlatch Mill," *Latah Legacy, The Quarterly Journal of the Latah County Historical Society*, 12:3 (1983), pp. 16-26.

51. Latah County Mining Notices, Book 1, p. 397; Book 2, pp. 69-77.

52. Like many other towns on the frontier, Palouse City experienced its "great fire" when most of the business district was destroyed. In the aftermath, many of the new business establishments were constructed of brick. A.J. Swarts originally purchased the W.F. Beyersdorf ranch on Meadow Creek. *Palouse Republic*, 12 Dec. 1917; *Bensel's Palouse City Directory for 1891*, p. 53.

53. Starner is today known as Hampton.

54. For Orcutt's change in occupation see *Palouse News*, 1 Jan. 1890. The quotation is from Kincaid, *Palouse in the Making*, p. 14. Also see *Palouse News*, January 8, 1892; May 2, 1892.

55. *Palouse News*, May 5, 1893; Northern Pacific Railroad Company, *The Fertile and Beautiful Palouse Country in Eastern Washington and Northern Idaho* (St. Paul, MN: The Northwest Magazine 1889), p. 24. In Palouse, J.G. Powers bought about $40,000 in gold dust per year, and J.H. Wiley bought about $15,000 annually. While these purchases did not constitute a legitimate claim for the upper Palouse Valley as a "minerals center," they showed that the Palouse River and its tributaries provided miners with enough gold mineral to give them a working wage after expenses.

56. *Palouse News*, June 30, 1893; August 11, 1893; August 25, 1893; Homer David, *Moscow at the Turn of the Century* (Moscow: Latah County Historical Society, 1979), pp. 20, 42, 80; Charles J. Munson, *Westward to Paradise* (Moscow: Latah County Historical Society, 1978), pp. 134-35; Hershiel Tribble, interviewed by Sam Schrager, July 16, 1973, Latah County Historical Society oral history collection; Milbert, "Men of Gold Creek," p. 3; Karen Broenneke, "The McConnell Mansion," *Latah Legacy, The Quarterly Journal of the Latah County Historical Society*, 9:4 (1980, esp. pp. 1-5). During this time, hard currency was so scarce that one Palouse City merchant sent his son to the mines in a light wagon to pay high prices for gold from the Palouse River miners.

57. *Palouse News*, 15 Sept. 1893; Latah County Mining Notices, Book 2, pp. 131-9; Whitman County Mining Claims, Record A, pp. 48-9, 88, 106-7; General Land Office Map, 10 May 1900, T42N R2W BM, R. Bonser, Report 2053.

58. Many of the early homesteading families, such as Chambers, Cochrane, Graham, Blake, Layton, and Hoskins had placer claims on nearby creeks. The most successful family enterprise grew out of John Truax's strike. The Taylors, from the Garfield-Farmington area, were partners in the Truax operation; their prospecting efforts led to the Mountain Gulch Mining and Milling Company. Frank Milbert, transcript, interviewed by Sam Schrager, June 19, 1975 and June 20, 1975, Latah County Historical Society oral history collection, pp. 84-6; Edna Johnson Butterfield, interviewed by Laura Schrager, October 11, 1973, Latah County Historical Society oral history collection; Tribble oral history interview.

59. Latah County Mining Notices, Book 2, p. 203.

60. E.S. Brents, along with John and Frank Maupin, located at the mouth of Poorman Creek; F.O. Slaght, Ed Graham, Racy Roberts, and George Henwood were on the Baby Grand Divide at the head of Slate Creek; W.G. Connor was at Quartz Gulch; and J.K. Truax was on the divide above Hoodoo Gulch. Latah County Mining Notices, Book 3, pp. 289, 293, 353-3, 367-9, 387; Book 4, p. 44.

61. Latah County Mining Notices, Book 3, pp. 295-6, 354, 390; Nez Perce County Deeds, Book 43, p. 485; Whitman County Mining Claims, Record A, pp. 48-9; Gilbert, *Historic Sketches*, pp. 18, 427. In addition to hearing mining news because of his stagecoach connection, Hemingway probably learned firsthand of Hoodoo strikes. The Muncys and their partners, who were active Hoodoo prospectors located one of their "hedge claims" on the Snake River directly across from Illia in April 1893.

62. To be sure, men like Jesse Bishop, J.C. Northrup, and the Taylor brothers should be counted among this group. When they came to the region they gradually developed their mineral and real estate interests.

63. Whitman County Articles of Incorporation, Book 1, articles 1, 15, 17, 99, 100, 139, 208, 240, 322, 412, 459.

64. Bockmier papers, folder 10. See also Milbert oral history transcript, pp. 16-7; Milbert, "Men of Gold Creek," p. 29; and Glen Gilder, transcript, interviewed by Sam Schrager, May 22, 1975, Latah County Historical Society oral history collection, p. 20.

65. The Blue River Mining Company had as principal prospectors W.J. and T.J. Demorest of Clarkston and Spokane.

66. Whitman County Articles of Incorporation, Book 1, article 99; Kincaid, *Palouse in the Making*, p. 18. For the impact of the Potlatch Lumber Company in the area, see Petersen, "Farewell to the Potlatch Mill," Petersen, "Life in a Company Town: Potlatch, Idaho," *Latah Legacy, The Quarterly Journal of the Latah County Historical Society*, 10:2 (1981), pp. 1-13.

67. Latah County Mining Notices, Book 3, pp. 595-604, 613-6; Charles G. Taylor to J.C. Lawrence, December 12, 1904, in John Craig Lawrence papers, Washington State University Manuscripts, Archives and Special Collections, Cage 1715; General Land Office Mineral Survey No. 2425, January 18, 1909.

68. There are no extant documents that show registrations by Northrup on Racy Roberts's original claims until 1904. There was no mining company incorporation by Northrup and his Palouse City associates until the organization of the Mizpah Copper Mining Company, Ltd., in 1906.

69. Hubbard, *Mineral Resources of Latah County*, p. 8; Wiley, "Gold Districts of the Northwest," p. 12; Latah County Mining Notices, Book 3, pp. 636-8; Book 4, pp. 43, 71-9; Whitman County Articles of Incorporation, Book 1, Article 139.

70. *Northwest Mining Truth*, December 15, 1916; January 3, 1920; Austin B. Clayton, "Copper Veins of the Mizpah Mine in the Hoodoo District Near Harvard, Idaho," unpublished master's thesis, University of Idaho, 1934, p. 2; Stanley Norman, *Northwest Mines Handbook* (Spokane: Northwest Mining Assn., 1918); Hubbard, *Mineral Resources of Latah County*, p. 7.

71. Miller, *Trees Grew Tall*, pp. 90, 111; Norman, *Northwest Mines Handbook*: Latah County Mining Notices, Book 4, pp. 131-8; Byers Sanderson, transcript, interviewed by Sam Schrager, January 23, 1976, Latah County Historical Society oral history collection, pp. 4-5, 18-9.

72. F.O. and J.O. Slaght, E.K. Parker, V.P. Wiesenthal, Pat Flynn, the Steffens brothers, E.S. Brents, A.J. Breeding, and C.W. Sanderson were among the many miners who embarked on such efforts, all being active along Poorman Creek and the North Fork Palouse River. Norman, *Northwest Miners Handbook*; T.A. Richard, *The Stamp Milling of Gold Ores* (New York: The Scientific Publishing Co., 1897), p. 265; Edgar K. Soper, "Brief Descriptions of the Mining Districts of Idaho and the Characters of the Ores," Washington State University Manuscripts, Archives, and Special Collections, VF 1021-231, p. 162; United States Geological Survey, 1912, p. 575.

73. Hubbard, *Mineral Resources of Latah County*, pp. 7-8; Milbert oral history transcript, p. 89; *Northwest Mining Truth,* 16 June 1932.

74. Hubbard, *Mineral Resources of Latah County*, pp. 10-11.

75. Milbert, "Men of Gold Creek," pp. 106-11.

76. Frank Milbert, personal communication with the author.

77. Doffner came to the Hoodoos in the late 1890s. After his death in the late 1930s, ownership of his claims was tied up in litigation until 1944. Freeberry was the last of the Hoodoo placer miners. He died in his cabin on Hoodoo Gulch in the mid-1960s. Latah County Probate Record, File 2671; Abe McGregor Goff, Moscow, personal communication with the author.

78. See, for example, the obituary for Stephen DeGrush, *Moscow Idaho Post*, February 12, 1914.

79. Among the most significant of these were Jim Lockridge, Frank Points, W.G. Connor, Cy Roberts, B. Norris Blake, Ed Graham, Stephen DeGrush, John English, C.W. Sanderson, Hans Lund, and the Chambers brothers.

80. John Wellington Finch, *Prospecting for Gold Ores*, Idaho Bureau of Mines and Geology Bulletin No. 36 (Moscow: University of Idaho, 1932), pp. 21-3; W.W. Staley, *Elementary Methods of Placer Mining*, Idaho Bureau of Mines and Geology Bulletin No. 35 (Moscow: University of Idaho, 1932), pp. 2-3; Eugene B. Wilson, *Hydraulic and Placer Mining* (New York: John Wiley & Sons, 1898), p. 21; Otis E. Young, Jr., *Western Mining* (Norman: University of Oklahoma Press, 1970). Little in the way of detailed descriptions of placer mining in the Hoodoo District has been found. Documentary evidence of the techniques used is limited to hydraulic structures, such as water ditches, located in mining claim notices, or brief seasonal progress reports to be found in contemporary newspapers. Autobiographical accounts and oral histories provide some insight for the later periods. The best evidence comes from investigating the material remains. The following, then, is a description of the placer technology suggested by those glimpses from documentary records, but it is based primarily on evidence from intensive archaeological surveys.

81. Livingston-Little, *Economic History of North Idaho*, p. 41; Merle Wells, *Rush to Idaho*, Idaho Bureau of Mines and Geology Pamphlet No. 19 (Moscow: University of Idaho, 1963); Young, *Western Mining*, pp. 58-60, 108-111; William S. Greever, *The Bonanza West: The Story of the Western Mining Rushes, 1848-1900* (Norman: University of Oklahoma Press, 1963), pp. 93-157; Rodman Paul, *California Gold: The Beginning of Mining in the Far West* (Cambridge: Harvard University Press, 1947), pp. 130-143,172-3;

Robert L. Kelley, "Forgotten Giant: The Hydraulic Gold Mining Industry in California," *Pacific Historical Review*, vol. 23, no. 4, pp. 343-56.

82. J. Ross Browne, *Reports on the Mineral Resources of the United States*, 1868, 40th Cong., 3rd Sess., H. Ex. Doc. No. 54, 1868, pp. 530-1; Young, *Western Mining*, p. 29; Lawrence K. Hodges, *Mining in Eastern and Central Washington* (Seattle: Shorey Book Store, 1970), pp. 187-92.

83. Lode deposits are mineral deposits found in solid rock.

84. Finch, *Prospecting for Gold Ores*, pp. 21-2.

85. Staley, *Elementary Methods of Placer Mining*, pp. 2-5; Philip S. Smith, "A Sketch of the Geography and Geology of Seward Peninsula," Water Supply Paper 314, United States Geological Survey, 1913, p. 39.

86. Young, *Western Mining*, pp. 22, 108-9; Wilson, *Hydraulic and Placer Mining*, pp. 21-4; C.C. Longridge, *Hydraulic Mining* (London: The Mining Journal, 1910), p. 181.

87. Young, *Western Mining*, p. 113; Wilson, *Hydraulic and Placer Mining*, p. 26.

88. Staley, *Elementary Methods of Placer Mining*, pp. 8-9; Gary D. Stumpf, *Gold Mining in Siskiyou County, 1850-1900*, Occasional paper no. 2 (Yreka, California: Siskiyou County Historical Society, 1979).

89. Staley, *Elementary Methods of Placer Mining*, pp. 10-11; Harold A. York, "The History of the Placer Mining Era in the the State of Idaho," unpublished master's thesis, University of Oregon, 1939, p. 44; Frank Milbert, personal interview with the author, 1980.

90. Virginia City *Montana Post*, April 29, 1865; Wilson, *Hydraulic and Placer Mining*, pp. 24-31.

91. Augustus J. Bowie, *A Practical Treatise on Hydraulic Mining in California* (New York: D. Van Nostrand, 1887), pp. 241-3; Frank Milbert, personal interview with the author, 1980.

92. Wilson, *Hydraulic and Placer Mining*, p. 34; Staley, *Elementary Methods of Placer Mining*, p. 12, Figure 10; T.F. Van Wagenen, *Manual of Hydraulic Mining* (New York: D. Van Nostrand, 1897).

93. Paul, *California Gold*, p. 114; Browne, *Reports of Mineral Resources of the United States, 1868*, p. 184; Wilson, *Hydraulic and Placer Mining*, pp. 648; Young, *Western Mining*, p. 122; Palouse Ranger District file 2540.

94. J. Ross Browne, *Reports of the Mineral Resources of the United States, 1867*, 40th. Cong., 2nd Sess., H. Ex. Doc. 202, p. 16; Browne, *Reports of Mineral Resources of the United States*, 1868, p. 180; Wilson, *Hydraulic and Placer Mining*, pp. 45-6; Bowie, *A Practical Treatise on Hydraulic Mining*, p. 135; Rossiter, W. Raymond, *Statistics of Mines and Mining*, 42nd Cong., 3rd Sess., H. Ex. Doc. 210, pp. 406-10.

95. Wilson, *Hydraulic and Placer Mining*, p. 45; Bowie, *A Practical Treatise on Hydraulic Mining*, p. 136; Frank Milbert, personal interview with the author, 1980.

96. Care of the ditch was so critical that miners in Alaska were known to have lined them with canvas. F.F. Henshaw and G.L. Parker, *Surface Water Supply of the Seward Peninsula, Alaska,* Water Supply Paper 314, U.S. Geological Survey, 1913, pp. 258-9; Milbert, "Men of Gold Creek," p. 4; Van Wagenen, *Manual of Hydraulic Mining*, p. 56.

97. Nez Perce County Mining Notices, Book P, p. 255; Nez Perce County Water Rights, Section D., XX, Book 1, pp. 18, 24-7; Browne, *Reports on the Mineral Resources of the United States, 1868*, p. 520.

98. Bowie, *A Practical Treatise on Hydraulic Mining in California*, pp. 93-4; Wilson, *Hydraulic and Placer Mining*, p. 75.

99. Raymond, *Statistics of Mines and Mining*, p. 408; Wilson, *Hydraulic and Placer Mining*, pp. 75-80.

100. Frank Milbert, personal interview with the author, 1980.

101. Bowie, *A Practical Treatise on Hydraulic Mining in California*, p. 82.

102. Wilson, *Hydraulic and Placer Mining*, p. 92; Alfred H. Brooks, *A Description of Methods of Placer Mining*, Water Supply Paper 314, U.S. Geological Survey, 1913, p. 286, Plate XIB.

103. Henry DeGroot, "Hydraulic Mining, parts II-VIII," *Mining and Scientific Press*, vol. 68, no. 1, Jan.-Feb., 1894; Bowie, *A Practical Treatise on Hydraulic Mining in California*, pp. 47-9; Karl Grove Gilbert, *Hydraulic Mining Debris in the Sierra Nevada*, Professional Paper 105, U.S. Geological Survey, 1917, pp. 104-7; Kelley, "Forgotten Giant;" Robert L. Kelley, "The Mining Debris Controversy in the Sacramento Valley," *Pacific Historical Review*, vol. 25, no. 4, 1956, pp. 331-46; Bull, "Palouse Valley History," p. 8; Emmett Utt, transcript, interview no. 1, Latah County Historical Society oral history collection, p. 30; Frank Milbert, personal interview with the author, 1980.

104. Raymond, *Statistics of Mines and Mining*, pp. 416-18 and Fig. 396a; Stumpf, *Gold Mining in Siskiyou County*, pp. 28-40; Wilson, *Hydraulic and Placer Mining*, pp. 69-72; Kelley, "Forgotten Giant," p. 355.

105. Bowie, *A Practical Treatise on Hydraulic Mining in California*, pp. 49, 162; Wilson, *Hydraulic and Placer Mining*, pp. 53, 58.

106. Wilson, *Hydraulic and Placer Mining*, p. 59.

107. Bowie, *A Practical Treatise on Hydraulic Mining in California*, pp. 82-3.

108. Milbert, transcript of interview, Latah County Historical Society oral history collection, pp. 99-100.

109. Browne, *Reports on the Mineral Resources of the United States, 1868*, p. 9; Wilson, *Hydraulic and Placer Mining*, p. 99.

110. Trimble, *Mining Advance into Inland Empire*, p. 227; Paul, *California Gold*, pp. 210-14; Charles Howard Shinn, "Land Laws of Mining Districts," *Johns Hopkins University Studies in Historical and Political Science*, Vol. II *Institutions and Economics*, No. XII, 1884, pp. 45-6, 61; Shinn, *Mining Camps: A Study in American Frontier Government* (New York: Charles Scribner's Sons, 1885) pp. 39, 44-5, 122-8.

111. Latah County Mining Notices, Book 2, p. 469. For a systematic study of authority conflicts between bureaucracies and various camp organizations see Janice Eleanor Nicholson, "Conflicts of Authority: An Analysis of Relations Among Authority Structures in the Nineteenth-Century Gold Rush Camps," unpublished Ph.D. dissertation, York University, Ottawa, Canada, 1973.

112. Trimble, *Mining Advance Into Inland Empire*, pp. 87-9; Browne, *Reports on Mineral Resources of the United States*, 1868, p. 530; Burcham, "Reminiscences of E.D. Pierce," p. 32; Charles B. Hunt, "Dating of Mining Camps with Tin Cans and Bottles," *GeoTimes* 3(8), 1959, pp. 8-10, 34.

113. Shinn, *Mining Camps*, pp. 281-4; Paul, *California Gold*, pp. 229-30, 234-5; Curtis H. Lindley, *Treatise on American Law Relating to Mines and Mineral Lands*, (San Francisco: Bancroft Whitney, 1897), pp. 61-2.

114. Latah County Mining Notices, Book 2, p. 117; Book, p. 612; Book 4, pp. 43, 45.

115. Latah County Mining Notices, Book 1, pp. 125, 132; Book 4, pp. 335-9; Nez Perce County Deeds, Book N, p. 209; Schell, *Ghost Towns and Live Ones*, pp. 3, 22, 34.

116. Robert E. Strahorn, *The Resources and Attractions of the Idaho Territory* (Boise: Territorial Legislature, 1881), p. 851; Milbert oral history transcript, p. 13.

117. Trimble, *Mining Advance Into Inland Empire*, pp. 45, 144.

118. Burcham, "Reminiscences of E.D. Pierce," p. 24; Milbert, "Men of Gold Creek," pp. 8-9; *Polk's Latah County Directory, 1905* (Seattle: R.L. Polk & Co.), p. 75; Ruby Canfield Wheeler, Latah County Historical Society oral history interview transcript, pp. 3, 6.

119. Bank of Palouse City depositors' ledger, 1888-1889, Cage 328, Washington State University Manuscripts, Archives and Special Collections, pp. 23, 66, 106, 201, 308, 351, 352; Palouse Businessmen's Association papers, Boomerang Museum, Palouse.

120. Burcham, "Reminiscences of E.D. Pierce," p. 135; Fred L. Israel, ed., *1897 Sears, Roebuck Catalogue* (New York: Chelsea House, 1968), p. 48; *Palouse Republic*, 9 May 1913.

121. Ira Kinman, Whitman County Historical Society oral history interview; Hershiel Tribble, Latah County Historical Society oral history interview; Glen Gilder, Latah County Historical Society oral history interviews; Verna Palmer Hardt, Latah County Historical Society oral history interview transcript, p. 2; Milbert, "Men of Gold Hill," p. 24.

122. General Land Office Map T42N, R2W, BM, R. Bosner Report 2053, 1899; George Nichols, Frank Herzog, and Glen Gilder, Latah County Historical Society oral history interviews; Edna Johnson Butterfield, Latah County Historical Society oral history interview.

123. Glen Gilder, Latah County Historical Society oral history interview no. 1, transcript, p. 18.

124. Byers Sanderson, Latah County Historical Society oral history interview no. 3, transcript pp. 27, 29; Milbert, "Men of Gold Hill," pp. 11, 20-23.

Index